**Property
Investment**

Property Investment

Principles and Practice of Portfolio Management

Martin Hoesli & Bryan D. MacGregor

Routledge
Taylor & Francis Group

LONDON AND NEW YORK

To Nanda Nanthakumaran

Just as we were finishing the final draft of this book, our friend and colleague
Nanda Nanthakumaran died suddenly and unexpectedly.
We would like to dedicate this book to his memory.

First published 2000 by Pearson Education Limited

Published 2013 by Routledge
2 Park Square, Milton Park, Abingdon, Oxon OX14 4RN
711 Third Avenue, New York, NY 10017, USA

Routledge is an imprint of the Taylor & Francis Group, an informa business

ISBN 13: 978-0-582-31612-6 (pbk)

British Library Cataloguing in Publication Data
A catalogue entry for this title is available from the British Library

Set by 35 in 10/12 Times and Helvetica

Contents

Part III
Property in a wider context

Preface

In comparison to the other main investment classes of shares and bonds, research on the property investment market has been more limited. However, such research has grown substantially in recent years in universities and in the private sector. A number of quality academic journals are now well established in the USA and the UK. More recently, international conferences in the USA, Europe, Asia and the Pacific Rim have developed, some aimed mainly at academics, some with a stronger professional focus. Applied property research has also developed substantially within the major institutional investors and in small specialist research companies. Many countries now have university courses in property, although in a few countries there has been a longer tradition of these.

Among the more important interests of this research have been the role of property in multi (mixed) asset portfolios, the measurement of property performance, the valuation (appraisal) process, the construction and management of property portfolios, modelling and forecasting rents and development, indirect property investment vehicles and international property investment. Despite the expansion of property research, general investment texts make little or no reference to property.

Property investments exhibit distinctive characteristics in comparison to other investment media, including fixed location, heterogeneity, high unit value and illiquidity. However, these characteristics should not justify property's separate and different treatment either in research or practical investment. Rather, they should inform the application of analytical tools from other markets to the investment management of property with other asset classes. This is particularly true as property markets become more international and as alternative property investment vehicles develop. Property market analysis also requires knowledge of the specialist area of urban economics but this needs to be integrated with an understanding of the capital markets.

The aim of this book is to draw together recent developments in property analysis from academic research and professional practice, and to apply these to the practical management of property portfolios. The focus is mainly on direct commercial property, but indirect vehicles and some aspects of residential property investment are also covered. The book is in three parts: the first covers the investment background, in particular, the characteristics of the various asset classes; the second presents the analyses necessary to develop a property portfolio strategy; and the third examines property in a wider context. At the end of each chapter, suggested further readings are provided. The book also contains an extensive list of cited references, where up-to-date international research output can be found.

Given the differences between property markets in different countries, no book could cover even the most important markets to the satisfaction of participants in those markets. Instead, we have tried to develop a general approach to the issues but, inevitably, given our experiences and the availability of material, there is a UK emphasis with strong European and US flavours.

This is primarily a textbook, targeted at later undergraduate years and graduate students. It requires some background in financial mathematics, statistics, economics and finance but, more importantly, it requires a willingness from the reader to tackle new material and not to be intimidated by equations. We hope it will be of value to the many courses world wide which now cover property investment. It should also be useful to finance, business and accountancy courses which seek to cover property. Although it is not a practice manual, we have tried to apply the more academic material to practical property portfolio management so that researchers and investment managers in practice should find material of interest and value.

Acknowledgements

This book was conceived at least ten years ago, so has had a lengthy gestation period. The delay in writing has certainly added to its quality as we have continued to develop our ideas and to learn from the work of others. Collaborators in research and consultancy, academic and professional colleagues, and participants at academic conferences and professional training courses have all helped us to broaden our knowledge and to develop our understanding of the field. Both undergraduate and graduate students have unknowingly been guinea-pigs as we have tested and developed the material.

Far too many people have contributed to this book, knowingly or unknowingly, for us to list them individually, particularly as errors or omissions would be inevitable. Instead we offer thanks collectively to colleagues and students, particularly at the Universities of Aberdeen and Geneva; staff, past and present, of the Property Research Team of Prudential Property Managers and Henderson Real Estate Strategy in London, and the Centre d'Information et de Formation Immobilières (CIFI) in Zurich; and participants at the research conferences of the American Real Estate and Urban Economics Association (AREUEA), the American Real Estate Society (ARES), the European Real Estate Society (ERES) and the Royal Institution of Chartered Surveyors (RICS).

The time and efforts were much appreciated of a number of people who commented on drafts of the whole book or individual chapters. Special thanks should go to Andrew Baum (Henderson Real Estate Strategy and University of Reading), Eva Cantoni (University of Geneva), Pierre-André Dumont (University of Geneva), Pat Hendershott (Universities of Aberdeen and Ohio), Catherine Jackson (University of Aberdeen), Colin Lizieri (University of Reading), Richard Shima (Abu Dhabi Investment Authority) and Aminah Yusof (University of Aberdeen). As ever, all remaining errors are entirely our responsibility.

While the writing of this book has been a highly motivating activity to its obsessive authors, the same could not be said for families, relatives and friends who have probably heard a little too much about it in the last two years. For their encouragement, patience and tolerance we are grateful. Particular thanks should go to Nicola MacGregor for becoming a book widow for a second time and to

Catriona and Eilidh MacGregor who by now probably think all men spend their evenings and weekends in their studies typing on computers.

We would also like to thank our publishers for patience and Chris Leeding, in particular, for his light-handed and supportive approach during the seemingly inevitable delays caused by overambitious timetables and competing commitments.

Finally, thanks are due to the British Council and the Swiss National Science Foundation and their British–Swiss research fund for supporting our collaborative work, much of which has found its way into this book.

Martin Hoesli & Bryan D. MacGregor
Aberdeen and Geneva, May 1999

Chapter 1

Introduction

1.1 The changing context for property investment

Property is one of the main investment assets, the others being shares and bonds. Shares (also known as stocks or equities) and property (also known as real estate) entail ownership rights and are known as *real* assets. In contrast, bonds are *financial* assets. Shares represent a claim to a proportion of a company and, so, are a form of joint ownership. Direct property investment involves the rights of ownership to a piece of land, typically with a building. Bonds are debt vehicles and a distinction can be made between government and corporate bonds. The importance of the two types varies substantially from country to country. In the UK, where they are known as *gilts*, government bonds are by far the more common; in the USA, company bonds represent one-third of the market; and in Switzerland, corporate bonds are more important.

Property comprises both *commercial* and *residential* investments. There are a number of different meanings of the term 'commercial'. In some countries it means offices and shops, with industrial as a separate category; in others, it also includes industrials; and in the USA, commercial refers to income producing (as opposed to owner occupied) property and includes offices, shops, industrials, rented apartment blocks and hotels. In this book, the term commercial property is taken to mean offices, shops and industrials and it is distinguished from residential property. Where clarity requires, 'property' is used to refer to the aggregate investment class and 'building' to refer to an individual property. Elsewhere, the term property is used in either an aggregate or individual sense which should be clear from the context.

The types of property which constitute the investment market and their financial characteristics vary from country to country. In the UK, the dominant form of property investment is commercial, while in other countries, such as the Netherlands and France, residential property is an important investment vehicle.

Property as an asset class exhibits distinctive characteristics, such as fixed location, heterogeneity, high unit value, illiquidity and, often, the use of valuations rather than prices for market information. There is also substantial government intervention, particularly in the housing sector. These factors and longer holding periods, partly as a result of lower liquidity, mean that, in terms of market turnover, shares and bonds are the predominant investment assets. However, in terms of value, property – including ownership by government, companies and individuals – represents by far the most important asset class. The relative importance of these owners, and of owner-occupation compared to investment ownership, vary substantially from country to country.

Property investment management has changed substantially in the last 20 years with its increasing integration with that of the other main investment classes. There have been two important consequences for the investment analysis of property. First, it has expanded from the selection of individual buildings to include a portfolio perspective in which portfolio structures are set relative to a benchmark and property is traded more frequently. Second, analysis has developed to consider property more explicitly in the context of the capital markets and the wider economy.

This integration has also made some of property's distinctive features more problematic. Dissatisfaction has developed about illiquidity, large lot size and the use of valuations to measure returns. This has been one factor in the development of indirect vehicles to gain exposure to the property market. The management of property investments has also been affected by the trends of globalisation in economic activity and investment. Property markets, particularly in major cities, have become increasingly international in terms of both occupiers and owners.

Thus, investors in property, particularly major insurance and pension companies (the *institutions*), have to make choices among the different investment media, and increasingly in different countries. These choices have to take into account the characteristics of the various asset classes and the linkages among them.

Data to analyse the property investment market are much more limited than in the share and bond markets. Aggregate data for property returns have been available since the late 1960s in the UK and 1978 in the USA, but in other countries such data are typically available for only a few years or not at all. This information is an important input into the comparison of property with other asset classes and for property portfolio construction. Its non-availability may help explain why property was for so long analysed and managed as a separate asset class and why most textbooks on investment ignore property.

A related explanation for the separation of property from other asset markets is the existence in many countries of a separate property investment profession. In the UK and the British Commonwealth, chartered surveying has until recently dominated investment analysis and buying and selling in the commercial property market. In the USA, the real estate profession is well-organised and several professional designations exist. In contrast, in mainland Europe, the property specialists are often architects or economists.

There are two models for the academic study of property. In the UK and the British Commonwealth, there is a separate education based on valuation, law, construction and urban economics, and typically separate from housing. Particularly in the UK, the study of finance has expanded in these courses to meet the demands of a changing market. In the USA and mainland Europe, it is more common to study property in the context of finance, often with urban economics and including housing.

The availability of sufficient time series for the analysis of property investment is an important explanation of the dominance of the USA and the UK in professional and academic property research. The top property journals covering investment research are produced in these countries: *Real Estate Economics, The Journal of Real Estate Finance and Economics* and the *The Journal of Real Estate Research* in the USA and the *Journal of Property Research* in the UK.

Professional practice, education and the academic study of property have developed substantially in the last 20 years. In most countries, property investment analysis requires consideration of the wider context of the capital markets and the economy, and research and education have developed to accommodate this. This book is a contribution to these developments.

1.2 The approach of the book

A number of general themes are pursued throughout this book: property investment is considered in a multi- or mixed-asset context; the investment management of property is viewed from a portfolio perspective; alternatives to investment in the direct domestic market are considered; and there is a strong applied, but theoretically rigorous, emphasis.

As it is necessary for investors to make choices between the different investment media, senior decision-makers require investment analyses for property which are comparable with those available for other investments. These decision-makers, with overall responsibility for investment decisions in the major investment institutions, are not property specialists but have backgrounds as actuaries, accountants and economists. Accordingly, they require analyses which use the vocabulary of investment and finance rather than the specialist terms that have developed in the property market. An approach is required which considers the other major asset classes, the linkages between assets and the linkages between the property market and the wider economy.

While property is different from the other main asset classes, this does not justify its being considered in isolation. The differences between the property market and other asset markets have been used as a justification for specialist knowledge and special and separate treatment of property. This book rejects such a stance and shows how techniques which are widely used in other asset markets can be modified and applied to property. To do so requires an understanding of these other markets and of the differences between them and the property market. Such an understanding should enable a more successful integration of property investment analysis with that prevalent in other markets.

The traditional focus for property investment has been the building specific level rather than the aggregate portfolio level. However, most investors hold portfolios of property and these have features different from a simple sum of their component parts. Thus, the main focus of the book is on the consideration of property in a portfolio perspective rather than at the individual building level. Larger investors have been developing this strategic approach to property portfolio management in which performance is judged against a market benchmark.

The portfolio perspective requires consideration of the drivers of property returns through links to the capital markets and the wider economy. The book develops these relationships and shows how they can be modelled.

The main focus is direct property, that is ownership of actual buildings, but indirect ways of gaining exposure to the property market are also considered. These have developed partly in response to concerns about the direct market, including illiquidity, large lot size and the use of valuations to measure returns. Gearing (leverage) to improve performance is covered in the context of indirect property investment vehicles, but not for direct property portfolios in which it is not an important factor. The growing importance of international property investment, both direct and indirect, as an alternative to domestic investment is considered.

Throughout, there is a strong applied focus, but it is soundly based in theory and techniques. The latest research findings are integrated into the discussions throughout the book and detailed further readings are provided.

1.3 The contents

The book is divided into three main parts. Part I (Chapters 2 to 4) covers background material which is essential for an understanding of the multi-asset context and the differences between property and the other asset classes. The main investment classes, including the concept of a risk-free rate of return, and the main types of investors are considered (Chapter 2). Also considered are the measurement of return and risk (Chapter 3) and performance indices (Chapter 4).

Part II (Chapters 5 to 9) presents the analyses necessary to develop a property portfolio strategy. First, a pricing framework for property is set out (Chapter 5). Next, the various inputs are considered: the expected income growth rate (Chapter 6); a risk premium (Chapter 7); and a depreciation rate (Chapter 8). Finally, much of the material is applied to the practical management of property portfolios (Chapter 9). Part III (Chapters 10 to 12) examines property in a wider context. This covers the role of property in multi-asset portfolios (Chapter 10), alternative ways of gaining exposure to the property market (Chapter 11) and international property investment (Chapter 12).

In order to understand the property market, it is essential to set it in the context of the other investment markets with which it competes for investors. In *Chapter 2* the characteristics of shares, bonds and property are discussed and compared. The links between these markets and from them to the wider economy are considered. Government bonds are introduced as an asset from which can be derived a risk-free rate of return. Other assets which bear risk must be priced to deliver the risk-free return plus a premium for risk. Finally, different types of investor and their investment requirements are examined.

Chapter 3 considers the measurement of return and risk. Measures of return are then discussed and it is shown how two types of investment yields, common to all asset classes, are linked. The conventional measure of risk (that is, the standard deviation) is also considered. Finally, the trade-off between return and risk, risk-adjusted returns and diversification of risk are examined.

To measure the return and risk at the asset class level, property indices are needed. The various ways of constructing property indices are the focus of *Chapter 4*. In the commercial property market, the lack of data on transaction prices has led to the construction of indices based on valuations of a sample of properties. Methods based on transaction data (averages of transaction prices, the hedonic method and the repeat sales method), which are mostly used in the housing sector, are presented. The measurement of returns from data on property company shares is also considered.

Chapter 5 considers the application of discounted cash flow (DCF) techniques to determine whether an investment is correctly priced. Central to the use of a pricing model is the notion that there may exist market inefficiencies, resulting in mispricing which can be exploited. The chapter first considers market efficiency and mispricing, particularly in the context of the property market. It next shows how to calculate the correct capitalisation rate and to compare it with the market rate and, equivalently, to calculate the expected return and compare it with the required return. This chapter also demonstrates how inputs from other markets can be used to assist with the assessment of property market mispricing.

Chapter 6 is concerned with the forecasting of property income, capitalisation rates and total returns. First, the relevant statistical tools required to build forecasting models are presented. Next, a simple model which links rents to the demand for, and supply of, property is estimated. It is then shown how this can be developed to produce a three-equation model in which demand and supply equations can be separately estimated. The simple model framework is also extended to the regional and town level. Finally, the forecasting of capitalisation rates and total returns is briefly discussed.

Chapter 7 deals with risk in a portfolio context. Modern Portfolio Theory (MPT), a theoretical framework for calculating risk and return when assets are considered in a portfolio, is considered first. Next, the Capital Asset Pricing Model (CAPM), which examines risk/return relationships in investments when the market is in equilibrium, is examined. Return and risk are also considered relative to a benchmark. Finally, the risk of not being able to meet liabilities is outlined.

Depreciation is an important feature of property assets and is considered in *Chapter 8*. First, definitions of depreciation based on its effect are presented. The causes of depreciation are then considered and used to classify depreciation. Finally, various methods of measuring the level of depreciation are discussed.

Chapter 9 applies the material presented in previous chapters to the practical construction and management of property portfolios. Active portfolio management and its application to property portfolios are considered first. The detailed issues are illustrated by an example which sets out considerations related to the broad structure of the portfolio and to the individual properties. Finally, practicalities involved in implementing the approach are outlined.

Chapter 10 considers the role of property in multi-asset portfolios. The role of commercial property, residential property and property company shares in diversifying mixed-asset portfolios is considered. The results from the various studies presented in this chapter are compared to the actual holdings of institutions, and some possible reasons for the observed differences are discussed.

Alternative ways of gaining exposure to the property market are reviewed in *Chapter 11*. These include indirect equity instruments (property company shares, other collective equity investment vehicles and unitisation), direct and indirect debt instruments and property derivatives.

Finally, *Chapter 12* examines whether investors should increase their range of investments by considering international property investments. The chapter begins with a review of the equilibrium relationships that should apply to product prices, interest rates and exchange rates, and which form the basis of international finance. The case for international diversification is made and the empirical evidence for shares, bonds, direct and indirect property is reviewed. Approaches to constructing an overseas portfolio and the associated problems are also discussed.

Each of the chapters sets out a rigorous analysis with up-to-date references from the academic literature and with detailed suggestions for further reading. Taken together they are intended to provide the theoretical basis and practical tools to assist with the management of a property portfolio in the context of a multi-asset portfolio.

Part I
Investment background

Chapter 2

Investments and investors

Introduction

Property is one of a number of types of investment available to investors.
Each of these investments has particular characteristics which are of varying
attraction to different categories of investors owing to their differing requirements.
Investments are initially supplied on a primary market between the 'issuer' and
the first purchaser. Thereafter, they are traded with third parties on a secondary
market. For each of the main investments, with the exception of property, there is
a central market.[1]

The function of investment markets is like that of all other markets in that they
bring together buyers and sellers. Investment markets are vehicles for transferring
savings into investments and they enable those with savings, the investors, to
acquire the investments which meet their requirements. In effect, investors swap
current capital for an income stream and future capital. Prices are set in these
markets to equate demand and supply.

The initial issuers vary from investment to investment. Companies issue *shares*
and *bonds* in order to raise capital to develop, to expand or to consolidate their
business activities; and governments issue bonds to enable them to borrow for
expenditure programmes. The *property* market is more complex as it comprises
separate development, investment and occupier markets. Developers provide
new property in order to generate profit for themselves. Investors who buy the
new property receive income from the occupiers in the form of rent and are able
to trade the property. Sometimes, however, the developer and investor are the
same entity.

In order to understand the property market, it is essential to set it in the context
of the other investment markets with which it competes for funds. This chapter
first sets out some fundamentals of investment in Section 2.2. It then identifies and
considers the characteristics of the principal types of investment in Section 2.3.
Next, in Section 2.4, there is a more detailed consideration of the characteristics
of the property market. In Section 2.5, investments are compared and categorised.
Different types of investor and their investment characteristics are examined in

Section 2.6. Finally, there is a summary and conclusions in Section 2.7. These discussions set an important context for detailed consideration of property investment in subsequent chapters.

Investment fundamentals

Nominal and real

The distinction between *nominal* and *real* is important in investment.[2] Nominal means in current money terms; real refers to the purchasing power of the money in terms of some base year. The difference is inflation.[3] Someone with £100 in 1991 would not have been able to buy the same as someone with £100 in 1971. The removal of inflation from nominal figures to produce real figures is known as deflating.

Example 2.1

Consider an investment which produces a constant nominal return of 10% during a time of variable inflation, as shown in Table 2.1. Calculate the real return.

Table 2.1► Deflating investment returns

Period	Inflation rate	Inflation index	Nominal return	Nominal return index	Real return index	Real return
	(1)	(2)	(3)	(4)	(5)	(6)
1	10%	1.10	10%	1.10	1.0000	0.00%
2	5%	1.16	10%	1.21	1.0476	4.76%
3	0%	1.16	10%	1.33	1.1524	10.00%
4	5%	1.21	10%	1.46	1.2073	4.76%
5	10%	1.33	10%	1.61	1.2073	0.00%

► Columns (1) and (3) provide the input data.
► The index in column (2) is calculated from (1): $(1 + 0.1) = 1.1$; $(1 + 0.1)(1 + 0.05) = 1.16$; $(1 + 0.10)(1 + 0.05)(1 + 0.00) = 1.16$; and so on.
► The index in column (4) is calculated from (3) in the same way.
► To obtain a real index, the nominal index (4) is divided by the inflation index (2).
► To obtain the real returns, the change in the index is divided by the starting value:
$(1.0476 - 1.0000)/1.0000 = 4.76\%$; $(1.1524 - 1.0476)/1.0476 = 10\%$; and so on.

Discounted cash flow techniques

The simplest way to consider any investment decision is to see it as the exchange of current capital for a future income stream, including a capital value at some time in the future. The current value of any investment depends on this income stream. Two fundamental issues then arise: the amount and timing of the income stream; and how the future values can be compared with the money required to buy it. This introduces the notions of the *time value of money* and *present value* of a future income stream.

The principle of the time value of money is that an amount of money receivable in the future is less valuable than the same amount receivable in the present. There are three reasons for this. First, money held now can be spent and utility obtained, whereas such consumption has to be forgone if the money is not to be received until some future date. Second, there may be a risk that all or part of the money may be lost in some way. For example, investment in a company risks the bankruptcy of the company. Third, if there is inflation, the purchasing power of the money will be reduced. Each of these factors requires compensation if the investment is to be made. This compensation can be considered as the return required by an investor to make an investment and so is known as the *required nominal return*. It may be expressed as:

$$R_N = RF_R + RP + i_e \qquad \qquad \text{◄Equation 2.1}$$

where: R_N is the required nominal return

RF_R is the real risk-free rate of return, the compensation for loss of liquidity

RP is the risk premium, the compensation for risk

i_e is the compensation for expected inflation.

It can be seen clearly that the higher the expected risk, the higher must be the required return. This relationship between risk and return is considered in more detail in Chapter 3.

If an amount of money, with a present value, PV, is to be invested for one period, it must earn a rate of return, R_N (the required nominal return). At the end of the period, the investor must expect to receive a total amount, A_1, equal to the original amount of the investment plus the required return on that investment. Thus:

$$A_1 = PV + (PV \times R_N)$$
$$= PV(1 + R_N)$$

Therefore:

$$PV = \frac{A_1}{1 + R_N} \qquad \qquad \text{◄Equation 2.2}$$

Equation 2.2 *discounts* the future amount A_1 and so converts it to a *present value*. More generally, an investment will generate income over several periods rather than over only one period. Each of these amounts can be converted into this common currency of present value. If the amount is receivable two periods into the future, it must be discounted over two periods, that is divided by $(1 + R_N)^2$; and if it is to be received i periods in the future, it must be discounted over i

periods, that is divided by $(1 + R_N)^i$. The individual amounts are added to produce the present value of the cash flows over n periods:

$$PV = \sum_{i=1}^{n} \frac{A_i}{(1 + R_N)^i}$$

◄Equation 2.3

where: PV is the present value

A_i is the amount receivable at the end of period i

R_N is the required nominal return (the discount rate)

$\sum_{i=1}^{n}$ is a mathematical symbol indicating that the amounts should be added for all values of i up to n. This is typically used in shorthand form as $\sum_{i=1}^{n}$.

The value of an investment, therefore, depends on the amount and timing of its cash flows and on the discount rate.[4]

Consider an investment that generates a cash flow of £100 at the end of each year for five years. If the required return is 10%, calculate the total present value of the investment. (See Table 2.2.)

Table 2.2► Present value calculation

Period	Future amount	Discount factor $(1 + 0.1)^i$	Present value
1	£100	$1.1^1 = 1.10$	£100/1.10 = £90.91
2	£100	$1.1^2 = 1.21$	£100/1.21 = £82.64
3	£100	$1.1^3 = 1.33$	£100/1.33 = £75.13
4	£100	$1.1^4 = 1.46$	£100/1.46 = £68.30
5	£100	$1.1^5 = 1.61$	£100/1.61 = £62.09
			Total = £379.08

Thus, if an investor were to pay £379.08 for the investment, s/he would receive an annual rate of return of 10%.

The total present value represents the worth of the investment to the investor based on his/her required return. If the investment trades at this price, it may be said to be fairly priced for the investor. If this price, P, is entered into the calculation, the *net present value* is zero, that is, the difference between the expected worth and the price is zero:

$$NPV = \sum_{i=1}^{n} \frac{A_i}{(1 + R)^i} - P = 0$$

◄Equation 2.4

Another way to consider the same issue is to calculate, from the cash flows and the market price, the return which will be delivered by the investment. This involves calculating the rate of return, R, which will equate the price to the present value of the cash flow, in other words, the net present value is zero. Thus, R, is such that:

$$\sum_{i=1}^{n} \frac{A_i}{(1 + R)^i} - P = 0 \qquad\qquad \blacktriangleleft\textbf{Equation 2.5}$$

This rate of return is known as the *internal rate of return* of an investment. It represents the *expected return* from an investment and is derived from the market price and the expected cash flows. The net present value and the internal rate of return can be used to derive two, equivalent, decision rules.[5]

▶ *The NPV rule*: If the net present value is greater than or equal to zero, an investor should buy.
▶ *The IRR rule*: If the internal rate of return is greater than the required return, an investor should buy.

The application of these rules to the management of property portfolios is considered in detail in Chapter 5. It is now possible to consider the cash flow characteristics of each of the main investment classes. These classes are conventional bonds, index-linked bonds, shares and property.

2.3 The financial characteristics of investments

Important financial characteristics of investments may be determined by consideration of their cash flows. The fundamental considerations are whether the income and the capital value are fixed or variable and, if variable, to what is the variation linked? Also of importance is how easy it is to sell the investments, that is, how liquid is the investment? The main investment classes are now considered.

Conventional bonds

A conventional bond, also known as a *fixed income security*, is a paper asset and is a form of *debt*. The bond is bought from the issuer for an amount known as the *par value*. In return the investor is entitled to a guaranteed nominal income, known as the *coupon*, each year and to have his/her money repaid at a set date, the *maturity* of the bond.[6] Bonds are usually issued for periods from 5 to 25 years. Some, known as undated, have no maturity, that is, the income is receivable in perpetuity and the initial purchase price is never repaid by the issuer. Bonds are described by the name of the issuer, the date of maturity and the coupon rate (the ratio of the coupon payment to the par value), for example, Treasury, 2005, 10%.

Conventional bonds can be divided into two categories: those issued by a government and those issued by others, mainly companies. In the UK and several other countries, such as France, Italy and Japan, the former is by far the more common. In the USA, company bonds represent about half of the value of government bonds. In Germany, the weights of government and corporate bonds are approximately equal, while in Austria and Switzerland, corporate bonds are more important.

Although the amount of income and the capital repayment are fixed, there is a risk that the issuer will default on these payments, for example, if the company becomes bankrupt. Bonds issued by the governments of countries such as those listed above are regarded as virtually riskless. Accordingly, they have a lower required return than bonds issued by others. In the UK, they are known as gilt-edged securities, or *gilts*.

Example 2.3

If an investor buys a five-year bond from an issuer for £100 with a 10% coupon rate and holds it to maturity, the cash flow will be as shown in Table 2.3. The internal rate of return for this bond is 10%.

Table 2.3► Conventional bond cash flow

Year	Type of cash flow	Amount of cash flow
0	Price	–£100
1	Income	£10
2	Income	£10
3	Income	£10
4	Income	£10
5	Income	£10
	Capital repayment	£100

Note: 'Period 0' in the conventional description of the start of a cash flow. For simplicity, it is assumed that the income is receivable in a single payment in arrears each year.

If a riskless bond is held to redemption, all the cash flows – the price, each income and the final capital value – are known exactly in nominal terms. It is, therefore, possible to calculate the nominal internal rate of return exactly. This rate of return to redemption is known as the *gross redemption yield*. This is a riskless nominal rate of return which can be used as a benchmark against which to assess the required return from risky investments.

However, bonds are traded up to redemption date on a secondary market. The secondary market is distinct from the primary market between the issuer and the purchaser. The initial investor can sell the investment on the secondary market to another investor prior to maturity; in turn this second investor can sell the bond to someone else in the secondary market, and so on. UK gilts are traded on the Stock Exchange, and the market is usually highly liquid. The price of the bond on the secondary market adjusts to deliver a competitive return. From Equation 2.1, it can be seen that one reason for a change in the required return is a change in inflation expectations (the real risk-free rate and the risk premium are considered below under index-linked gilts). Consider Example 2.4.

Example 2.4

Consider a 10-year gilt, initially purchased for £100, a coupon of £10 (and so a coupon rate of 10%) and five years to maturity. Suppose that a rise in expected inflation since the issue of this gilt has resulted in a rise in the required return, and a new, five-year gilt is being offered with a 12% interest rate and a coupon of £12. At what price should the first gilt trade so that the gross redemption yields are the same?

The solution of the correct price is the amount which will provide an IRR of 12%. This can be calculated by trial and error using a spreadsheet. In order to deliver a return of 12%, the price of the first gilt must fall to £92.79.

Table 2.4► Price adjustment in the gilt market

Period	Gilt 1	Gilt 2
0	£?	−£100
1	£10	£12
2	£10	£12
3	£10	£12
4	£10	£12
5	£110	£112
IRR	12%	12%

Thus, if inflation expectations change, the market price of the gilt will change. Conventional gilts are affected by inflation in another way. If actual inflation is other than expected, the real value of the income and the par value will be different from that expected. Thus, even if held to redemption, although the nominal IRR is fixed, *unexpected* inflation will mean a change in the real IRR. It can be said that conventional gilts are *prone to inflation*.

However, gilt returns are not linked directly to economic growth. They are not affected by changes in expected growth, or differences between actual and expected growth, unless these result in changes in inflation or expected inflation.

Index-linked gilts

Index-linked gilts (ILGs) are a special type of UK government bond. They were first introduced in 1981 as an inflation-free investment for pension funds but have been available to all investors since 1982. Only a few other countries, including Canada and the USA, have such investments. ILGs provide income and a redemption payment which are fixed in *real* rather than *nominal* terms. This is done by fixing the real income and redemption payment and then adding on inflation that has already occurred.[7]

Table 2.5 sets out a simplified calculation of the cash flows to be received from an ILG. The guaranteed real cash flows are multiplied by the inflation index to produce the nominal cash flow. If the ILG is held to redemption, the investment provides protection against inflation and the real IRR, that is the gross redemption yield, is guaranteed. The ILG, therefore, provides a real risk-free rate that can be used as a benchmark against which to assess other investments.

Table 2.5► Index-linked gilts

Period	Real cash flow	Inflation rate	Inflation index	Nominal cash flow
0	−£100		1.00	
1	£5	5%	1.05	£5.25
2	£5	2%	1.07	£5.36
3	£5	2%	1.09	£5.46
4	£5	5%	1.15	£5.74
5	£105	10%	1.26	£132.48
IRR	5.0%			9.9%

The ILG may be traded on a liquid secondary market in which prices adjust to ensure the delivery of a competitive real return. Therefore, if it is not held to redemption, the resale value will be unknown in advance and the real IRR will be unknown. The nominal income and IRR, even if held to redemption, are unknown as these vary with the unknown future rates of inflation.

Conventional and index-linked gilts are important in that their gross redemption yields provide, respectively, nominal and real risk-free rates of return. Recall that Equation 2.1 was:

$$R_N = RF_R + RP + i_e$$

Thus, the required return from an investment comprises the index-linked gilt yield, a premium for the risk of the investment and compensation for expected inflation. This can be applied to a conventional gilt. If the gilt market is correctly priced, the required return will be the same as the expected return, so its gross redemption yield will equal its required return. If real returns are important to an investor, the only risk for a conventional gilt is that the estimate of expected inflation is incorrect. This means that an inflation risk premium is required. The two bond yields may thus be compared:

$$RF_N = RF_R + iRP + i_e$$ ◄Equation 2.6

where: RF_N is the nominal risk-free rate
RF_R is the real risk-free rate
iRP is the inflation risk premium
i_e is the rate of expected inflation.

These risk-free rates can be used to analyse other investments which carry risk, and to build up their required returns (see Chapter 5). It is also possible to deduce the market's expectations of inflation if an assumption is made of the inflation risk premium.

Suppose the gross redemption yields on two gilts, conventional and index-linked, both with redemption dates of 2024, are 8% and 4%, respectively. What is the market's expectation of average inflation over the 25 years from 1999 to 2024?

Using Equation 2.6:

$$RF_N = RF_R + iRP + i_e$$

gives:

$$8 = 4 + iRP + i_e$$

Now suppose that the inflation risk premium is 0.5%, then:

$$8 = 4 + 0.5 + i_e$$
$$i_e = 3.5$$

Note, however, that the inflation risk premium is a variable. In periods when inflation is expected to be stable, it will be lower than when inflation is expected to be more volatile. Estimating a value for the inflation premium is a matter of judgement rather than precision.

The nominal redemption yield on index-linked gilts is linked directly to the rate of inflation as this is added to the fixed real income. However, like conventional gilts, ILG returns are not directly linked to economic growth.

Ordinary shares

The following discussion is restricted to *ordinary* shares as other types of shares are relatively unimportant. A share is a paper asset which, as the name suggests, carries with it a share in the capital and the income of the company and a share in the management of the company through voting rights proportional to the number of shares held. It is an *equity* investment, in contrast to bonds which are *debt* investments.

The income, the *dividend*, is paid twice yearly (interim and final) in the UK and is not guaranteed in either real or nominal terms; it depends on the profitability of the company and on the policy of its directors. Profits can be used for investment or for the payment of dividends. If profits are insufficient to provide an acceptable dividend, reserves or borrowing may be used. Thus, although there is a link between profits and dividends, it is not one to one.

The profits of the company depend on the state of the economy and the success of the company's business. If the economy is booming, profits and dividends should be higher than if the economy is in slump. In the longer term, real profits and dividends should grow in line with real levels of economic activity as productivity increases. Accordingly, they should grow in both real and nominal terms.

Shares have no redemption date and are traded on a secondary market, so the capital value is not guaranteed in either real or nominal terms. The Stock Exchange is the market where shares are traded. It is a highly liquid market.[8] As shown above, the value of an investment depends on the amount and timing of its cash flows and on the required return used to discount them. If the economic outlook improves, expected income will increase and so will the value of a share. If the real rate of interest and the ILG yield rises, other things being equal, the share discount rate will rise and the capital value of the expected income flow will fall.

The impact on share prices of a change in inflation expectations is more complicated and depends on the ability of a company to pass on inflationary rises in input costs to consumers (see Hoesli, MacGregor, Matysiak and Nanthakumaran, 1997). If all inflationary rises could be passed through to customers without affecting sales, there should be no effect on real dividends or on share prices.[9]

From the above, it is clear that the internal rate of return of a share is unknown and has to be estimated from its anticipated cash flows. Shares, therefore, involve risk and so the required return includes a risk premium, usually thought to be in the order of 2–3% in the UK.

Property

The last of the main investment classes is property, the main concern of this text. Its cash flow features are considered here, and in the next section the discussion is extended to other important features of property which distinguish it from other investments. Unlike bonds and shares, which are paper investments, the main property investments considered in this book involve the purchase of an actual property. This form of investment is known as direct property investment and is distinct from indirect property investments which are paper assets backed by property. Indirect property investment is considered in Chapter 11.

The types of property which constitute the investment market and their financial characteristics vary from country to country (see also Chapter 12). In the UK, the dominant form of investment in property (as opposed to ownership by occupiers) is in commercial property – that is, shops, offices and industrial and warehousing.[10] As of 1996, for instance, it was estimated that UK pension funds and insurance companies held 40% of their property holdings in offices, 45% in shops, 14% in industrials and 1% in other types of property such as hotels or forests (Dumortier, 1997).

In other countries, rental housing is an important form of property investment. Commercial property in the USA is estimated to comprise 67% of industrials, offices and shops, 28% of multi-family residential housing and 5% of hotels (Miles *et al.*, 1994). US pension funds, for instance, in 1996 held 16% of their property holdings in rental housing, 28% in offices, 27% in shops, 13% in industrials and 16% in other types of property. The importance of residential properties in the property portfolios of investors in some other countries is even larger. French insurance companies held 46% of their property portfolio in housing in 1996. Swiss pension funds hold 74% of the property component of their portfolio in apartment buildings, while this is as high as 84% for Swiss real estate mutual funds (Dumortier, 1997; Schärer, 1997).

In the UK, leases are usually for 15–25 years with rent reviewed periodically, typically every five years. Unlike the other main investment classes for which income is received twice yearly in arrears, rents are paid quarterly in advance.[11] Between reviews, property income is like that of a conventional bond, fixed in nominal terms and so prone to inflation. The reviews are 'upwards-only', that is, if prevailing market rents have risen, the rent will rise, but if they have fallen, the rent remains the same. This means that UK property returns are linked to levels of economic activity and so property has equity features, but without the risk of a fall in nominal income.

As outlined in Chapter 12, lease terms vary quite substantially across countries. In the USA, office leases are typically for five or ten years. The tenant pays a fixed base rent plus a proportionate share of the increase in building expenses. Base rent can be fixed for the whole lease term, increase at stipulated intervals or increase with inflation. Retail leases usually include a percentage rent clause under which the tenant has to pay a portion of annual gross sales turnover above a stipulated level.[12] Such leases exist also for retail property in France, where the percentage of sales which has to be paid to the landlord varies between 5 and 7%. The French commercial leases are usually for nine years with three-year break options in the tenant's favour.[13] Rents are indexed either annually or at the end of every third year, usually by the changes in the cost of building construction. In Germany rents are usually indexed to the cost-of-living index (Gelbtuch *et al.*, 1997).

The shorter the length of the lease or, if they exist, the period between rent reviews, the more quickly will rents adjust to the real economy and to inflation. In some countries, automatic adjustments, based on inflation, during the length of the lease provide further protection against inflation. In the UK, the five-yearly rent review is 'upwards only' so rent does not fall in nominal terms during a lease term. In all cases, over the long term, cash flows should be more or less in line with economic activity and property should have similar characteristics to shares. This is particularly true for retail property when percentage clauses exist. As is the case for shares, property should also provide long-term inflation protection.

The default risk of a property investment depends on the quality of the tenant, in the same way that the default risk of a bond depends on the issuer. It has traditionally been argued that property has a more secure income than shares as rent has to be paid, even if the company occupying the building is making a loss, while there is no such obligation to pay a dividend to holders of shares. Further, if the occupier goes bankrupt, it is possible to find another tenant, whereas if a company goes bankrupt, the shareholders lose part or all of their money. While true for shares of the quality held by major investors, this can be overstated. Moreover, these investments are held in diversified portfolios, so the impact of one or two bankruptcies on the overall portfolio is small.

Property, like shares, has no maturity and is traded on a secondary market.[14] Accordingly, the capital value may rise and fall in both nominal and real terms depending on actual cash flows, on future expectations of growth and inflation, and on the discount rate.

2.4 The characteristics of property investments

This section considers a number of other important features of property. These are: heterogeneity and fixed location; unit value; borrowing; the long-term nature of the holding; management; the development cycle and supply; depreciation; government intervention; psychic income; price determination; illiquidity; and classification of property types.

Heterogeneity and fixed location

All ordinary shares in a company are identical and all bonds of a particular issue are identical. In contrast, properties are *heterogeneous*, that is, each property is unique. They vary by location, use, size of plot, size of building, age, construction, maintenance and tenant.

The great importance of location in property investments stems from their immobility. Factors external to the property, such as access to infrastructure and quality of the area, fundamentally affect its value. However, the importance of location varies by use. It is crucial for retail units and values can vary substantially from one end of a shopping street to another, depending on pedestrian flows; but is much less important for offices in the same office park or industrial units on the same industrial estate.

In the same way that the financial strength of the issuer influences the market value of a bond, the financial standing of the tenant is an important factor in the value of a property. In both cases a contract with the government would be regarded as the most valuable. Thus, identical properties beside each other on the same business park could have different values depending on the tenant.

One consequence of heterogeneity is that, while it is possible for two investors to have identical holdings in shares and bonds, it is not possible to do so in property (see Chapters 4 and 7).

The importance of heterogeneity can be overstated and it has been used to argue in favour of detailed consideration of individual properties to the exclusion of a strategic overview. Although each property is different, there are systematic factors which affect the returns of all properties of a particular type. Thus, all shops would be affected by changes in consumer expenditure and all properties in a town would be affected by the closure of a major employer (see Chapter 6). Further, an increase in the real discount rate would cause all properties to fall in value.

Unit value

The unit value of a property holding is very much larger than in other markets: a few pounds will buy a share and £100 will buy a gilt, but even the smallest investment is likely to cost several hundred thousand pounds. It is, therefore, impossible for small investors to enter the market, and difficult for even significant investors to construct diversified property portfolios. This has important consequences for the way in which particular investors should gain exposure to the property market. In effect, it means that smaller investors should not invest in direct property but should hold units in an indirect vehicle (see Chapters 10 and 11). Securitisation and time share are means of reducing the unit value. The former is well established in some markets such as the USA, while the latter is a feature only of the holiday property market.

Borrowing

One important consequence of the high unit value of property is that borrowing is important. This is particularly true of property development which is generally undertaken using short-term finance. It is less true of direct investment in prime property in the UK as the market is dominated by major investors. However, such investors may provide commercial mortgages. This used to be a favoured way of gaining exposure to the property market (see Chapter 11). There may be tax advantages in having an investment partly financed by debt.

The long-term nature of the holding

Land is, generally speaking, indestructible and buildings have a long economic life expectancy, so it is seen as a long-term investment. Another important factor is that it is a real asset with long-term real returns linked to the economy. However, more active property portfolio management has led to more frequent trading (see Chapter 9). Thus, while a portfolio of properties may be seen as a long-term holding, individual properties may be bought and sold to generate short-term performance (net of transaction costs) in a competitive fund management market. In this sense, a property portfolio differs from a shares portfolio only in the frequency and scale of trading.

Management

Ownership of a bond can be regarded as a passive form of investment, that is, no management input is required beyond the decisions to buy and sell. Ownership of a share is also typically a passive investment. Although it carries with it voting *rights* (but no obligations), these are unlikely to be influential unless the percentage holding in the company is significant. Even then, the influence is over the broad strategic management of the company and the appointment of directors rather than day-to-day management. For both bonds and shares, price is determined on the secondary market and is beyond the control of an individual investor.[15]

Investment in property is rather different. Typically, advice is required not only on buying and selling, as it is for all forms of investment, but also on day-to-day management because the owner has significant management obligations. These include maintenance, rent collection, rent reviews and lease negotiations. Although they can be subcontracted (for example, to firms of surveyors), this means additional costs compared to other investments. These management obligations vary depending on the type of property: a high street shop requires less than a large shopping centre.

Some of the obligations can be seen as opportunities. Properties can be refurbished or redeveloped, and adjoining sites can be purchased to release extra value. These decisions can be timed according to market conditions to have maximum impact on value. Management of leases and rent reviews, for example in a shopping centre, can also affect income and thus value. Individual shop units can have rent reviews set to different dates or leases of different lengths so as always to have market evidence for subsequent review. In these and other ways, a property owner can add to the value of his/her property investment in a way that is not possible in the share and bond markets.

The development cycle and supply

One specific example of management opportunities is the creation of new property investments. Unlike other markets, it is possible for an investor to create his/her own investments. However, this is a lengthy and often risky process which requires land acquisition; financing; planning permission; a lengthy construction period; and a search for tenants. It is easier for industrials and more difficult for retail and offices. It is, therefore, difficult to increase supply quickly to meet new demand, which itself is difficult to forecast. This creates additional risk. Typically, increased demand triggers development activity but, by the time the new development is ready, demand may have begun to fall back. The consequence is fluctuations in the supply of new developments with consequences for property prices and so property returns (see Ball *et al.*, 1998 for a full discussion of property cycles).

Depreciation

A further distinctive feature of property is the management of depreciation. Depreciation comprises two parts. The first is deterioration over time as a result of wear and tear and the effects of elements. This results in a decline in the income-earning capacity of an investment, over time, when compared to an identical new investment. The second part is obsolescence which is related to changes in building technology and in the functional requirements of property occupiers. A common example is 1960's/70's office blocks with inadequate floor-to-ceiling heights for false floors and ceilings to accommodate electrical cables for modern office equipment or with concrete pillars preventing flexible open plan spaces (see Chapter 8).

However, the assets of a company also deteriorate and its capital equipment and products become obsolete. The fundamental difference is one of management. For a company these are the problems for the managers rather than the owners of the shares, whereas for property it is the responsibility of the owner.

Government intervention

The fixed nature of property, its relationship with neighbourhood and environmental quality, its importance in economic enterprises and the quality of life of households have all led to substantial government intervention in the property market. Examples include planning and environmental controls, building regulations, rent controls in housing, subsidised industrial and housing rents, development incentives in particular in run-down areas and restrictions on ownership. These add to the management obligations. In some countries, capital gains and transfer taxes increase the costs of trading.

Psychic income[16]

It has been suggested that psychological factors play an important factor in decisions to buy a property (see Baum and Crosby, 1995). In essence, it is argued that buyers are prepared to pay more than is justified by a rational analysis of the future income stream, for example, for offices in a prominent location or of an award-winning design. However, caution is required with this interpretation. If these psychological factors play an important role for purchasers, there is no reason to suppose that occupiers are not similarly affected and so are prepared to pay a higher rent. If this is the case, the higher purchase price is justified.[17] An opposite effect is likely to occur for properties of poor design, high management inputs and in unattractive locations. Such properties may trade below their worth. Thus, there may exist opportunities for the informed investor to exploit mispricing (see Chapter 5). If it is long term, an investor can benefit by receiving more income than required; if short term, there are opportunities for trading profits.

Price determination

Price in the property market is not determined by the interaction of many sellers and many buyers for a homogeneous investment in a central market, such as a share in a company or a gilt of a particular issue, to produce a market clearing price. There is limited information available on transaction prices as there is no single trading market, no central price listing for property, and the volume of transactions is relatively low. Heterogeneity also means that information on one transaction does not provide an exact proxy for the likely selling price of another property. Price, therefore, is the product of negotiation between the seller and one or more buyers for unique properties in local markets. This requires local knowledge, typically obtained from professional valuers, and so results in costs (see Chapter 5).

One important consequence of the shortage of transactions information is that it is difficult to track property market movements. This is fundamentally different from the shares and bond markets where transaction data are readily available. Instead, movements in the property market are often tracked by valuations, that is, estimates of likely selling prices rather than actual prices. This has led to mistrust among investment professionals of the information available on property market returns. It may also lead to market mispricing. These issues are considered in depth in Chapters 4 and 5.

Illiquidity

Property is an illiquid investment and buying and selling it is costly and time-consuming. In comparison to other investments, property is sold less frequently. There are a number of reasons for illiquidity, including: the large unit size; the uniqueness of each individual property; the complexities of the legal interests; the need for physical surveys; the absence of a single trading market to match buyers and sellers; and the difficulties of agreeing a price. Trading and capital gains taxes further limit liquidity. Although the gilt and main share markets are liquid, this is not true for all shares and bonds. There are other share markets, and shares are classified according to how often they are traded, although these low liquidity shares would not be a significant part of an institutional portfolio.

Classification of property types

While bonds can be classified by issuer and issue, and shares by individual company, there are broader groupings which seek to combine bonds or shares which are likely to have similar price movements because of systematic factors affecting them. Gilts are classified by time to redemption and shares by the industrial sector of the company.

The conventional way to classify property is by sector (office, retail, industrial, housing and so on) and by geographical region as these different classes are considered to have different economic drivers. The retail sector is driven by retail sales; industrials by manufacturing output; offices by business services output; housing by personal income; and so on. The regional sector markets are driven by the regional economies which have different economic structures. However, the regional division is not uncontroversial (see Hoesli, Lizieri and MacGregor, 1997). There is an argument for the geographical classification to be based on classifications of town by size or function, although no standard classification has emerged in the UK. Changing requirements for space, such as the combination of office and manufacturing uses and the expanding importance of warehousing

using industrial type properties, has also led to a blurring of sector definitions (see Chapters 6 and 9 for fuller discussions).

2.5 Comparing investments

From the above discussion, it is possible to summarise the financial characteristics of the main investment classes and to compare them.

Market size

Table 2.6 shows the market capitalisations of the main UK investment classes. The share market is by far the largest while the conventional gilt market and property markets (including owner-occupation) are of similar size, and the index-linked market is by far the smallest. However, the estimate of the value of all commercial property is somewhat speculative.

In the USA fixed income securities represent a much larger proportion of the main asset classes with $9554bn as compared to $4658bn in publicly traded securities and $3989bn in commercial property (Miles *et al.*, 1994).

Investment categories

The price of an investment is simply the present value of the expected income stream discounted at the appropriate rate to take account of the risk of the investment. The way in which price responds to changes in expectations of economic growth and inflation provides an important way of categorising investments. One fundamental distinction is between equity (or real) investments and debt (or money) investments. The former are generally linked to the real economy (levels of economic activity), and provide protection against inflation; the latter are linked to the money economy (inflation and interest rates) and are prone to inflation.[18]

Shares are an equity investment with a right to a share in the success of a company and their fortunes are linked to the real economy. This link means that, over a reasonable holding period, shares should provide protection against inflation. Conventional bonds are a debt investment with income fixed in nominal terms, so are not directly linked to the real economy.[19] Thus, if there is a rise in inflation expectations, the nominal discount rate will rise and the price will fall, so bonds are prone to inflation. Index-linked gilts are a hybrid investment. Their real

Table 2.6► Market capitalisation of the main investment classes	
Investment class	Market capitalisation
UK share market	£1234bn
Conventional gilts	£264bn
Index-linked gilts	£54bn
Commercial property (excluding owner-occupation)	£120bn
Commercial property (including owner-occupation)	£280bn

Source: Investment Property Databank (IPD, 1997), Datastream, Greenwells (see Ball *et al.*, 1998).

income is fixed so is not linked directly to the real economy, but they were designed specifically to provide protection against inflation.

UK property contains features of shares and conventional bonds: the importance of each depends on the prevailing state of the economy. In periods of high growth, property owners benefit from the economic growth, as do owners of shares; in periods of low growth or declines in output, its 'upwards-only rent reviews' protect investors from falls in nominal income and give property stronger bond features.[20] Over the longer term, property, like shares, should provide protection against inflation.

Property in other countries has similar features to shares. Leases are usually shorter than in the UK and can be adjusted both upwards and downwards, making it possible for contract rents to track market rents more closely. In several countries, retail rents are adjusted upwards if sales are above an agreed level. The link between economic activity and cash flows should thus be quite strong in many countries. Lease contracts in most countries provide for rental adjustments within the lease duration. These adjustments are often made according to some index of price changes, such as an inflation index or a construction cost index.

Thus, property is linked to real economic activity and inflation, so in most countries it should provide the investor with protection against inflation, at least in the long term. Hamelink *et al.* (1997), for instance, report that with a seven-year time horizon in the USA and 13-year horizon in the UK there is a 95% probability of achieving a positive real return. For shares, these figures are 16 years in the USA and four years in the UK, respectively. For bonds, 12 years and 19 years, respectively, are needed.

Table 2.7 summarises the financial characteristics of the main investment classes.

2.6 The main investors and their investment requirements

The institutions

Each investment class has different features which make it more or less attractive to different investors with different requirements. Although some wealthy individuals are active in investment markets, the dominant investors are organisations. These include insurance companies, pension funds, investment trusts, unit trusts, savings banks, building societies and property companies.[21] Collectively, the first four are known as the *institutions*. These organisations are the intermediaries through which most people invest. In effect, individuals have transferred responsibility for their investment strategies to the institutions.

The growth in the institutions in the UK has, in part, been the consequence of government regulation, such as compulsory insurance, the tax advantages of life assurance policies, and obligations, since 1975, on employers to provide pension schemes (Rutterford, 1993). The institutions control between one-half and two-thirds of each of the main investment markets and the proportion has been increasing. The insurance companies and pension funds account for around 90%

Table 2.7► Financial characteristics of investments

	Conventional bonds	Index-linked gilts	Equities	UK property
Nominal income fixed or variable	Fixed	Variable	Variable	Fixed between reviews, otherwise variable upwards only
Real income fixed or variable	Variable	Fixed	Variable	Variable
Nominal capital value fixed or variable	Fixed if held to redemption, otherwise variable	Variable	Variable	Variable
Real capital value fixed or variable	Variable	Fixed if held to redemption, otherwise variable	Variable	Variable
Nominal expected return fixed or variable	Fixed, if held to redemption, otherwise variable.	Variable	Variable	Variable
Real expected return fixed or variable	Variable	Fixed, if held to redemption, otherwise variable.	Variable	Variable
Security of income	Secure if UK government, otherwise depends on issuer	Secure (issued by UK government)	Depends on company	Depends on tenant
Liquidity	Gilts liquid, but general bonds depend on issuer and market size	Liquid	Liquid	Illiquid
Links to the economy	Real return depends on inflation	Nominal return depends on inflation	Linked to economic growth and inflation	Linked to economic growth and inflation

of institutional investment holdings, split roughly between them. Thus, to understand the investment markets, it is necessary to understand something about the requirements of these major investors.

Investor requirements

In the UK, there are three major types of institutional investor: general insurance funds, life assurance funds and pension funds. The two types of insurance funds must, by law, be managed separately. Each of the three has different liabilities and so has different requirements of investments. While each type of fund invests in all of the main assets, the balance of their portfolios reflects their particular requirements. A general picture for the UK can be seen in Table 2.8.

However, such percentages vary quite substantially from one country to another (Dumortier, 1997). Pension funds and insurance companies in most countries have much higher allocations to fixed-income securities than is the case

Table 2.8► Investors and investments, UK, 1991

	Pension funds	General insurance	Life assurance
Cash	7.1	15.7	6.8
Gilts	9.2	14.6	13.5
<5 years	0.4	7.6	0.8
5–15 years	4.2	6.7	8.0
>15 years	1.4	0.3	3.1
ILGs	3.2	0.0	1.6
Shares	49.0	26.5	40.9
Property	8.7	10.2	15.2
Loans and mortgages	0.1	5.0	3.2
Overseas	18.0	15.9	10.8
Other	7.9	12.0	9.6

Source: CSO, Business Monitor (reported in Rutterford, 1993).
Note: 'Other' includes unit trust units and agents' balances.

in the UK. In Japan and Denmark, for instance, the proportion of assets invested in bonds is 64% and 67%, respectively. Accordingly, the proportion of assets in shares varies quite substantially across countries. Canadian institutional investors hold 60% of their assets in shares but Swiss institutions hold only 19%. The Swiss institutions on the other hand hold 16% of their assets in property, while US institutions only hold 3% of their assets in property.

Part of the explanation of the differences lies in regulatory frameworks, which may set minimum or maximum limits for some asset classes, and in the relative sizes of the domestic markets, which may limit the opportunities available to an investor. Although the following discussion is UK-oriented, the basic principles can be applied to other countries.

General insurance covers items such as car, house and contents insurance. It is typically for periods of one year at a time. It is a reasonably straightforward task to predict the number of car crashes, car thefts, burglaries and house fires. The premium paid by the policy-holders is based on the probability of a claim. It is rather more difficult to predict hurricanes and droughts, hence the problems faced by these companies after the storms of October 1987 and the structural damage to foundations after two dry summers in the south of England in the early 1990s. General insurance requires cash flow for claims and so these funds invest more in liquid assets with fixed income such as short-term cash deposits and short-dated gilts rather than those with variable income but long-term capital growth, such as equities. However, they are in a competitive business and so need to diversify into investments classes to achieve competitive returns and to keep premiums low.

Life assurance typically combines two parts: life insurance, that is, if the policy holder dies, his/her beneficiary will receive a capital sum; and a long-term savings policy, often related to the profits of the investment. Accordingly, they require longer term holdings with competitive returns. However, as the amount to be paid on death is often fixed in nominal terms, conventional bonds are also attractive.

There are several types of *pension fund*. Some, termed occupational pension funds, are for particular occupations, such as university lecturers; others are funds in which individuals buy units. The latter have been growing in importance with the expansion of private pensions. The former tend to provide pensions based on final salary while the latter tend to provide pensions based on the value of the units. Final salaries are likely to grow in line with the real economy, so the funds require investments, such as equities and property which have long-term income and capital growth linked to the real economy. The money purchase schemes operate in a competitive market and so require competitive long-term returns, requiring significant investment in real assets.

For the purposes of this discussion, the most important distinction is between immature funds and mature funds. Pension funds are designed to grow to maturity, at which point their net income available for investment should fall to zero.[22] Immature funds are net receivers of income, that is members' contributions and investment income exceed current pension payments. If a fund is continually gaining new members and losing others through death, it will have longer term liabilities and so require longer term real investments, that is, shares and property. Conventional gilts, with fixed nominal income and a redemption value fixed in nominal terms, are less attractive to such funds. However, index-linked gilts, with protection against inflation, have attractions. If the fund is a closed fund, with no new members, it will have increasing requirements to pay pensions from investment income and the sale of assets rather than from members' contributions. Its requirements, then, are shorter term.

For smaller funds, the large lot size of property and the inability to construct diversified portfolios make it an unattractive investment. As pension funds are typically smaller than life funds, on average they hold lower proportions of property. These issues are considered in more detail in Chapters 4, 7 and 9.

A further factor of relevance to investment strategies is the tax position of investors. For example, exemption from capital gains tax, but a liability to taxes on investment income, would make investments with low income returns but high expected capital gains attractive propositions.

2.7 Summary and conclusions

The central theme of this chapter has been that property is only one of a number of investments which investors can choose to meet their investment requirements. Different types of investor, with different requirements, find each investment class to have varying degrees of attractiveness. An understanding of property investment, therefore, requires an understanding of these competing investment markets and the requirements of investors. Two related issues are the linkages among these markets and between the markets and the wider economy.

Investors are able to select from conventional gilts, index-linked gilts, shares and property. The first two are debt or money investments with returns linked to the money economy, that is, inflation and interest rates. Importantly, if held to redemption, conventional gilts provide a nominal risk-free rate of return and index-linked gilts provide a real risk-free rate. These are crucial building blocks in an understanding of the required return on other risky investments.

Shares are an equity investment with returns linked to both the money economy and the real economy, that is, the level of production. UK property has hybrid features: in times of economic growth, upward-only rent reviews allow an equity type return linked to the real economy; in times of negative growth, fixed nominal income between reviews and upward-only reviews give it conventional bond features. In other countries the features of property resemble more closely those of shares.

Property was also identified as having a number of features with important consequences for its investment characteristics and management. The most significant of these are:

▸ Large unit value and heterogeneity mean that smaller investors of all types are unable to construct properly diversified property portfolios.
▸ Property investment carries with it management rights and obligations which mean costs but may also provide additional returns.
▸ Property returns are linked to the development cycle. Increases in the level of economic activity trigger development booms which usually lead to over supply and to a dampening of rental growth.
▸ The reliance on valuation for tracking property market movements has led to scepticism among investors about the merits of property as an investment.
▸ There may be mispricing which can be exploited.
▸ Property cannot be traded easily and quickly.

The main investors in all the investment markets are the institutions: general insurance funds, life assurance funds and pension funds. These have varying liability structures and so varying requirements for money and equity investments, although each invests in portfolios of all of the main investment classes.

Thus, property investment must be seen in the context of other investments, investor requirements and the wider economy. It must also be seen in a portfolio context. These issues are explored in subsequent chapters but, first, it is necessary to consider the two most basic concepts in investment: return and risk.

Notes

1. The stock market is for shares and bonds and there is a money market for short-term cash deposits. There are also markets in futures and options, currencies, commodities and overseas financial instruments.
2. For example, the *volume* of economic output is a measure of real output while the *value* is a measure of nominal output. Instead of using real and nominal, the expressions constant price and current price are often used. *Constant* price means real price while *current* price means nominal price.
3. Strictly, the difference could also be because of *deflation*. This is, in effect, negative inflation and the purchasing power rises rather than falls. It is a very rare event in most economies.
4. Note that there is no reason why the discount rate should not vary from period to period.
5. More generally, for example when choosing between projects, the NPV is a superior method and the IRR is fraught with difficulties. For a discussion, see Lumby (1988).

6. Strictly, the annual income is paid every six months in arrears in two equal amounts.
7. In the UK, the compensation for inflation, as measured by the Retail Price Index (RPI), is lagged by eight months. Therefore, there is a small risk that actual inflation will be higher or lower than that for which compensation is paid.
8. Shares not traded on the main Stock Exchange are less liquid.
9. The effect of inflation also depends on the debt structure of the company and on tax laws. See Hendershott (1981) for a discussion.
10. Whereas in the UK commercial property encompasses shops, office and industrial and warehousing, in the USA it also includes residential properties held for investment purposes. These residential properties mainly include multi-family residential buildings and hotels.
11. This makes difficult the comparison of capitalisation rates (the ratio of income to price), as the present value of an income stream receivable quarterly in advance is greater than the same annual amounts receivable twice yearly in arrears. This issue is discussed in Chapter 5.
12. Typically this portion is between 1% and 9%.
13. Turnover (overage) rents, upward-only rent reviews and break clauses are examples of options in property leases. The valuation of these options has begun to be addressed in the property literature (see Grenadier, 1995; Ward et al., 1998).
14. The exception is a leasehold property where the property reverts back to the issuer of the lease. In this case, the capital value to the leaseholder at maturity is zero.
15. The exception would be a large investor with a significant holding in a company. In this case, a decision to sell a substantial number of shares would move the market against the seller. This, of course, would be unlikely to occur as it would not be in the interests of the investor.
16. In US textbooks this characteristic is often referred to as pride of ownership.
17. Note, that for owner-occupied property, this simply means that the imputed rent (as opposed to the actual rent) is higher.
18. The term 'protection' is used deliberately to distinguish it from hedging. It means a real return over a 'reasonable holding period'. Strictly an inflation hedge has returns which move exactly in line with actual inflation from period to period. The academic evidence suggests that shares are not, generally, an inflation hedge: the UK is one of a handful of countries where there is any evidence to support the proposition. Evidence for property being an inflation hedge is a little more mixed and its inflation-hedging features may vary according to the economic environment.
19. There is, of course, a link between the real and money economies. For example, a period of above-average economic growth is likely to lead to a rise in interest rates in an effort to dampen demand and inflation.
20. This must be qualified by the risk of default, either of the issuer of the bond or the tenant of the property.
21. Property companies are considered in detail in Chapter 11.
22. There are statutory requirements which mean that assets must exceed liabilities.

Further reading

Baum, A. E. and Crosby, N. (1995) *Property investment appraisal*, Routledge, London (second edition): Chapters 1 and 2.
Brett, M. (1990) *Property and money*, Estates Gazette, London: Chapters 1 and 42.
Brueggeman, W. B. and Fisher, J. D. (1997) *Real estate finance and investments*, Irwin, Chicago (tenth edition): Chapters 3, 8 and 10.

Corgel, J. B., Smith, H. C. and Ling, D. C. (1998) *Real estate perspectives: an introduction to real estate*, Irwin/McGraw-Hill, Boston (third edition): Chapters 2, 3 and 9.

Darlow, C. (ed.) (1983) *Valuation and investment appraisal*, Estates Gazette, London: Chapters 1, 2 and 3.

Fabozzi, F. J. and Modigliani, F. (1992) *Capital markets: institutions and instruments*, Prentice Hall International Editions, Englewood Cliffs (NJ): Chapters 1, 8 and 15.

Fraser, W. D. (1993) *Principles of property investment and pricing*, Macmillan, London (second edition): Chapter 21.

Howells, P. G. A. and Bain, K. (1990) *Financial markets and institutions*, Longman, London: most chapters.

McIntosh, A. P. J. and Sykes, S. (1984) *A guide to institutional property investment*, Macmillan, London: Chapters 1, 2 and 3.

Ross, S. A., Westerfield, R. W. and Jaffe, J. (1996) *Corporate finance*, Irwin, Chicago (fourth edition): Chapters 4 and 5.

Rutterford, J. (1993) *Introduction to stock exchange investment*, Macmillan, London (second edition): Chapter 12.

Sharpe, W. F., Alexander, G. J. and Bailey, J. V. (1995) *Investments*, Prentice Hall International, Inc., Englewood Cliffs (NJ) (fifth edition): Chapters 14, 15 and 17.

Wurtzebach, C. H. and Miles, M. E. (1991) *Modern real estate*, John Wiley & Sons, Inc., New York (fourth edition): Chapters 1 and 21.

Chapter 3

The measurement of return and risk

3.1 Introduction

Return and risk are central to investment theory and practice. As outlined in Chapter 2, the higher the expected risk of an investment, the higher is the required return. The measurement of return and risk should, therefore, be fundamental to all investment activity although, in practice, the focus is much more on return. For example, fund managers advertise their successes in achieving above-average returns but do not include measures of the risk taken to achieve these.

Both historical (*ex post*) and expected (*ex ante*) returns and risk are important and are considered in this chapter. The measures of historical return are used to analyse risk and to estimate depreciation. They are used to compare the property market as a whole with the other main investment markets and to compare different property types and geographical areas.[1] They are also used to compare individual fund performance with competitors or to the market, and to identify the causes of differential performance. Historical data are also used to model the future, on the assumption that 'the future will be like the past'. Such analyses include building forecasting models and portfolio construction models (see Chapters 6 and 7, respectively).

Return and risk, whether historical or expected, can be measured at the level of the individual investment, an investment portfolio or the market as a whole. This chapter considers general measures which are suitable for all of these; the next chapter considers index construction to measure market returns.

Section 3.2 considers measures of return. It first considers simple measures of income return, capital return and total return. Next it sets out more complex measures that are used either to deal with expenditures during a period, or to combine returns for a number of periods. These are the money-weighted rate of return, the internal rate of return and the time-weighted rate of return.

Section 3.3 then considers investment yields and shows how two of these are closely linked to measures of return. Two others are particular to the property market because of its pattern of cash flows. The section then focuses on the use of yields to produce expected returns rather than historical returns. In particular, it shows how the income yield can be used to deduce market expectations about the future values of key variables. This aspect is developed more fully in Chapter 5.

The focus then moves to risk and Section 3.4 sets out the conventional measure of risk – namely, the standard deviation of returns. The use of historical data rather than expectations information is highlighted. Consideration is given to which factors cause actual returns to be other than those expected. Certainty equivalent cash flows and scenario analysis are then outlined. Next, the risk – return trade-off, risk-adjusted returns and diversification of risk are examined. Finally, other dimensions of risk are dealt with. Section 3.5 provides a summary and conclusions.

3.2 The measurement of return

Income, capital and total returns

The *income return* is the net income received over the measurement period divided by the value at the beginning of the period:

$$IR_t = \frac{NI_t}{CV_{t-1}}$$
◄Equation 3.1

where: IR_t is the income return for period t
NI_t is the net income received during period t
CV_{t-1} is the capital value at the start of period t (the end of period $t - 1$).

For property, the income should be the rent actually *received* rather than receivable and it should be net of any outgoings or expenses.

The *capital return* is the increase in the capital value over the measurement period divided by the value at the beginning of the period:

$$CR_t = \frac{CV_t - CV_{t-1}}{CV_{t-1}}$$
◄Equation 3.2

where: CR_t is the capital return in period t
CV_t is the capital value at the end of period t
CV_{t-1} is the capital value at the end of period $t - 1$.

The *total return* is simply the sum of the income and capital returns:

$$TR_t = \frac{(CV_t - CV_{t-1}) + NI_t}{CV_{t-1}}$$
◄Equation 3.3

where: TR_t is the total return. This definition of *total return* is often termed the *money-weighted rate of return* (MWRR).[2] It takes into account both the magnitude and the timing of cash flows. Equivalent measures may be calculated in the share and gilt markets. In the share market, it is called the *holding period return* and is defined as:

$$HPR_t = \frac{(P_t - P_{t-1}) + D_t}{P_{t-1}}$$

◄Equation 3.4

where: HPR_t is the holding period return in period t
P_t is the price at the end of period t
P_{t-1} is the price at the end of period $t-1$
D_t is the dividend received during period t.

A specific problem for property (in this and in all measures of return) is the need to use valuations to estimate capital values whereas, for shares and gilts, actual market prices can be used. If the valuations are inaccurate, the measures of return are also inaccurate. This issue is considered in detail in Chapter 4.

The MWRR and the IRR

Equation 3.3 can be rewritten as:

$$\left[\frac{CV_t}{1 + TR_t} + \frac{NI_t}{1 + TR_t}\right] - CV_{t-1} = 0$$

◄Equation 3.5

where: CV_t is the capital value at the end of period t
CV_{t-1} is the capital value at the end of period $t-1$
NI_t is the net income received
TR_t is the total return (measured by the MWRR).

In other words, if the investment were bought for CV_{t-1}, were worth CV_t after one period and income of NI_t were received during the period, then TR_t would be the *internal rate of return*. This assumes that the income is received in a single payment at the end of the period. For other timings of the cash flows, the MWRR (as defined in Equation 3.3) is only an approximation to the IRR: the latter takes into account the timing of cash flows more accurately.[3]

The timing of cash flows

The above are simple measures of return in which it is assumed that all income is received at the end of the period. However, income from gilts is received twice a year, in arrears, in two equal parts; and for shares in two, usually unequal, parts. For UK property, the income is received quarterly in advance. This means the present values are different as shown in Example 3.1.

Example 3.1 ►

Consider three investments, each bought for £1000 and with an annual income of £100. The first has a single payment at the end of the year; the second has two equal payments in arrears; and the third has quarterly payments in advance. The present values of these income streams discounted at 10% are, as shown in Table 3.1, respectively, £90.91, £93.13 and £96.52.

Table 3.1► The present value of income with different payment patterns

End of quarter	Payment	PV @ 10%	Payment	PV @ 10%	Payment	PV @ 10%
0					£25	£25.00
1					£25	£24.41
2			£50	£47.67	£25	£23.84
3					£25	£23.28
4	£100	£90.91	£50	£45.45		
Total		£90.91		£93.13		£96.52

Note: Columns need not add to totals due to rounding.

It is also possible to calculate the future values of these incomes at the end of the year by multiplying, rather than dividing, by $(1 + R)^t$. In essence, this assumes that the income can be invested and interest will accrue until the end of the year. Again using a rate of 10%, the future values are, as shown in Table 3.2, respectively, £100, £102.44 and £106.18. This gives effective income returns of 10%, 10.244% and 10.618%. The MWRR (as defined in Equation 3.3) does not take account of this but the IRR does.

Table 3.2► The future value of income with different payment patterns

End of quarter	Payment	FV @ 10%	Payment	FV @ 10%	Payment	FV @ 10%
0					£25	£27.50
1					£25	£26.85
2			£50	£52.44	£25	£26.22
3					£25	£25.60
4	£100	£100.00	£50	£50.00		
Total		£100.00		£102.44		£106.18

Expenditures

A further complication in calculating returns comes from expenditures on the investment during the period. An assumption of no costs during the period is valid for gilts and shares and may be valid for a single property for some years, but not for a portfolio of properties: There are likely to be injections of capital for maintenance, refurbishments and developments. This pattern of expenditures can be complex. To overcome the problem, a variant on the MWRR formula of Equation 3.3 has been developed:

$$TR_t = \frac{(CV_t - CV_{t-1} - C_t) + NI_t}{CV_{t-1} + kC_t}$$

◄Equation 3.6

where: TR_t is the total return for period t
CV_t is the capital value at the end of period t
CV_{t-1} is the capital value at the end of period $t - 1$
C_t is the capital expenditure during period t
NI_t is the net income during period t
k is the fraction of the period over which the capital expenditure was applied.

In theory, this could be calculated for a portfolio:

$$TR_{P_t} = \frac{\sum\limits_{i=1}^{n} [(CV_{it} - CV_{it-1} - C_{it}) + NI_{it}]}{\sum\limits_{i=1}^{n} CV_{it-1} + k_i C_{it}}$$

◄Equation 3.7

where: TR_{P_t} is the portfolio return during period t

CV_{it} is the capital value of the ith property at the end of period t

CV_{it-1} is the capital value of the ith property at the end of period $t-1$

C_{it} is the capital expenditure on the ith property

NI_{it} is the net income from the ith property

k_i is the fraction of the period during which the capital expenditure was applied on the ith property.

In this formulation, the capital expenditure is considered to reduce the change in value over the period. The return received must be based on the *total* capital employed during the period, that is the initial value for the whole period *plus* the capital expenditure for a fraction of the period. If $k = 1$, the expenditure is at the beginning of the measurement period. In practice, it is typically assumed that all expenditure takes place at the mid point of the year, so $k = \frac{1}{2}$.

An alternative way to conceptualise the problem is that the capital expenditure is paid from income and so should *only* appear on the numerator and not in the denominator. This would hold if the expenditure were less than the income, which would be true for a portfolio but not necessarily for an individual property. If it were not, borrowing would be required. The formulation is:

$$TR_t = \frac{(CV_t - CV_{t-1}) + (NI_t - C_t)}{CV_{t-1}}$$

◄Equation 3.8

Note that, for any given inputs, this version will result in a higher measured return than Equation 3.6. As with Equation 3.3, this can be rewritten to show that TR_t is the IRR if the income is received and the expenditure incurred at the end of the period. More complex variants are used in the USA (see Young *et al.*, 1995).

Time-weighted rate of return

The MWRR can be considered to be a measure of return for a single period. When returns over several periods are being considered, it may be necessary to consider the impact of the timing of cash injections on fund performance. Example 3.2 shows the crucial impact of timing.

Consider two funds to which new capital is injected at different times. In the case of Fund 1, it is injected just before the market booms, whereas for Fund 2, an identical amount is injected just before the market falls. The consequence is that the final value of Fund 1 is £2m more than that of Fund 2. However, both sets of managers produced the same returns from their portfolios if the capital injections are excluded from the analysis. In this case the appropriate measure is the time-weighted rate of return (TWRR).

Table 3.3► The impact of timing of cash injections

Period	Fund 1				Fund 2			
	Start value	Return	End value	New capital	Start value	Return	End value	New capital
1	£100m	5%	£105m	£10m	£100m	5%	£105m	£0m
2	£115m	20%	£138m	£0m	£105m	20%	£126m	£10m
3	£138m	−10%	£124.2m		£136m	−10%	£122.4m	

The TWRR calculates the return up to the period of the capital injection and then 'chain-links' or compounds these separate period returns to calculate the average return over a longer period.

$$\text{TWRR} = [\Pi(1 + TR_i)]^{1/t} - 1 \qquad\qquad \text{◄Equation 3.9}$$

where: TR_i is the total return in period i

Π is the mathematical symbol, pi, indicating that the individual values of $(1 + R_i)$ have to be multiplied

t is the number of periods.

The TWRR measures the individual returns for each sub-period and so is not influenced by the volume or timing of capital cash flows. For example, in the analysis in Example 3.2, the rate, but not the amount, of return on the £10m available to Fund 2 during period 2 is taken into account. Accordingly, it is an appropriate measure for investments such as quoted unitised funds which do not have control over capital injection and removal. If, however, the fund manager has control over the timing of the capital injection, a different measure, such as the IRR, is required.

Consider the funds with cash flows as set out in Table 3.4. Both funds invest £100m at the start. Fund 1 produces a 50% return (+£50m) followed by 0%; and Fund 2 produces 0% followed by 50%. The TWRRs are the same as both funds have the same combination of annual returns, but the IRRs are different because the cash flows are different. In this case, the IRR is more appropriate as it takes into account the money available for re-investment in period 2.

Table 3.4► Time-weighted and internal rates of return

Period	Fund 1		Fund 2	
	Return	Cash flow	Return	Cash flow
0		−£100m		−£100m
1	50%	£50m	0%	£0m
2^Note	0%	£100m	50%	£150m
IRR		28.08%		22.47%
TWRR	22.47%		22.47%	

Note: The cash flow includes return of the original capital.

Example 3.4 shows how the three methods differ when fixed income and expenditure are included in the calculation.

Consider the cash flows from an investment as set out in Table 3.5. It is initially bought for £1m, then income of £75,000 is received in each of three years and expenditure of £250,000 is incurred in year 2. At the end of the three years it is worth £1.375m.

Note the following features of the analysis:

► The year by year MWRRs using $k = 0$ and $k = \frac{1}{2}$ differ in year 2 because of the expenditure.
► The two MWRR calculations of the annual returns produce different average annual TWRRs for the three-year period.
► The MWRR using $k = 0$ produces the same result as the year by year IRR. However, over the three years, the average return differs: the IRR is 8.39% and the TWRR is 8.61%. In general, the IRR takes into account the timing of cash flows more accurately.

Table 3.5► Money-weighted, time-weighted and internal rates of return

Cash flows

Period	Capital value	Income	Expenditure	Net cash flow
0	£1 000 000			−£1 000 000
1	£900 000	£75 000	£0	£75 000
2	£1 250 000	£75 000	£250 000	−£175 000
3	£1 300 000	£75 000	£0	£1 375 000
				IRR = 8.39%

Return calculations

Period	IR	CR	MWRR ($k = 0$)	MWRR ($k = \frac{1}{2}$)
1	7.50%	−10.00%	−2.50%	−2.50%
2	8.33%	11.11%	19.44%	17.07%
3	6.00%	4.00%	10.00%	10.00%
			TWRR = 8.61%	TWRR = 7.88%

The above example illustrates the need to use the most appropriate measure of return and for the measurement to be standardised if comparison of funds is to have any meaning. Some of the problems are particularly acute in the property market but others, such as how best to measure managers' performance, are common to all investments. These are important issues as less than one percentage point could mean a substantial difference in fund ranking in performance tables. In the next section, attention is turned to yields as measures of return, in particular what can be deduced from yields about market expectations.

The term 'yield' is used to describe a measure of return and there are substantial similarities between commonly used yield measures and the return measures discussed in the previous section. Accordingly, they can also be used as measures of historical return. However, in this section, the focus is on yields as measures of expected return. There are two general types of yield in investment. One refers to a measure of total return and the other to a measure of initial income return. Although the names differ from market to market, the basic concepts are the same.

Total return

Total return is measured by the internal rate of return (IRR). For gilts, held to redemption, it is called the *gross redemption yield* and its expected value can be calculated accurately for any dated gilts because all the magnitudes and timings of the cash flows (the purchase price, the coupon and the par value) are known.

(a) Consider a gilt bought at the end of 1998 for £90 with 10 years to redemption and a 10% coupon. Calculate the gross redemption yield.
(b) Consider a gilt bought at the end of 1998 for £120 with 10 years to redemption and a 10% coupon. Calculate the gross redemption yield.

Table 3.6▶ Gilt gross redemption yields

Cash flow	Year										
	1998	1999	2000	2001	2002	2003	2004	2005	2006	2007	2008
(a)	−£90	£10	£10	£10	£10	£10	£10	£10	£10	£10	£110
(b)	−£120	£10	£10	£10	£10	£10	£10	£10	£10	£10	£110

The gross redemption yields are the internal rates of return of the above cash flow. Note that the final cash flow includes the coupon (£10) and the par value (£100). For (a) the IRR is 11.75%, for (b) it is 7.13%.

In the share market, as shown in Equation 3.4, the standard measure of return is the holding period return (HPR):

$$HPR = \frac{(P_1 - P_0) + D_1}{P_0}$$

where: P_0 is the price
P_1 is the value at the end of the holding period
D_1 is the dividend.

As shown above, for income received annually in arrears, this is the IRR. While it is easy to calculate historically, it is difficult to predict. The future income stream is not known, nor is the capital value in the future: only the current price and past dividends are known for certain. Accordingly, it requires significant research and should be subject to sensitivity analyses using ranges of values for the inputs (see Section 3.4).

Other measures are used to help give an indication of future returns on shares. These are used partly because it is difficult to estimate the future income stream of a share. The *earnings per share* (EPS) is simply the amount of income for each share. The *price to earnings ratio* (P/E ratio) is the ratio of current price to the latest earning per share. The P/E ratio is, therefore, similar to the inverse of the dividend yield (see below), except that it considers earnings and not dividends paid. The *dividend payout ratio* (DPR) is the ratio of dividend per share to the earnings per share. The *dividend cover* is the reciprocal of the DPR and is the ratio of earnings to dividend. If dividend cover is unity, a downturn in earnings would probably force a cut in dividend or the use of reserves to maintain the dividend. If the dividend cover is greater than 1, the company could maintain dividends if earnings were to fall.

In the property market the measure of total return is called the *equated yield* and is estimated using estimated changes in rental value and in capital value. It should be calculated by an explicit consideration of the future income stream. Like its share equivalent it requires research and a range of values should be estimated using a range of values for the inputs. It is little used in practice and can be misused as it may be easy to manipulate the inputs to get the answer required.

Income yields

Income yields measure the relationship between the current income and the current capital value. For gilts, the initial yield is the coupon/current market price. Gilts have a fixed income, a fixed redemption value, but a variable market price depending on discount rates. The higher the discount rate, the lower is the present value of an income stream. The price of a gilt adjusts according to the discount rate: as the discount rate rises, price falls; and as the discount rate falls, prices rise. The income yield in the gilt market is known variously as the *interest yield*, *running yield* or *flat yield*.

For shares, the income yield is known as the *dividend yield*. The current value is the market price but the income, the dividend, is not known in advance and so the figures used in calculating the dividend yield refer to last year's known interim and final dividends rather than the unknown dividends for the current year. Thus, it is calculated as D_0/P_0 rather than D_1/P_0. If it is available, the yield is quoted with the latest interim dividend to indicate whether it is greater than that of the previous year.

For property, the income yield is known as the *initial yield* or the *all risks yield* (ARY) and it is the ratio of rental income (net of outgoings) to current value or price. In most cases, as price is unknown, it has to be estimated by valuation. The inverse of the income yield in the property market is known as the *years' purchase* (YP) and is equivalent to the present value of £1 received annually in arrears in perpetuity and discounted at the initial yield rate. It is analogous to the P/E ratio in the share market.

Property-specific yields

The pattern of income for UK property, that is, income fixed in nominal terms for five years between rent reviews, has led to the use of two property-specific yield measures: the *yield on reversion* (or reversionary yield) and the *equivalent yield*. Until the next review, the rent from a property is fixed but the open market rent will adjust to supply and demand. At most times, particularly with the impact of inflation, the open market rental value will be above the contract rent. The yield on reversion is defined as (net) estimated rental value (ERV) (as opposed to contract rent) divided by current value or price. Reversionary potential is defined as open market rental value divided by contract rent.

The equivalent yield is defined as the internal rate of return of the cash flow from the property, assuming a rise to ERV at the next review but with no further rental growth. It will always lie between the initial yield and the yield on reversion.

Income yields and market expectations

The income yield and the equated yield, while called different things in different investment markets, are common concepts. The former is a simple measure of initial income return while the latter is a measure of total return. The current value of an investment is the expected future income stream discounted at an appropriate rate to take account of its risk. Thus, the ratio of current income to current capital value, the income yield, provides information which can be used to deduce market expectations of income growth and risk. These issues are considered in detail in Chapter 5 but can be illustrated here with a simple example.

Consider the investments in Table 3.7.

Table 3.7▶ Deducing market expectations from income yields

Asset	Expected income					Cash flow
	Year 1	Year 2	Year 3	Year 4	Year 5	
A	10	10	10	10	10	Uncertain
B	10	10	10	10	10	Certain
C	10	15	20	25	30	Certain
		-2	-2	-2	-2	
D	10	15	20	25	30	Certain
		-1	-1	-1	-1	
E	10	15	20	25	30	Certain

Although assets A and B have the same expected cash flow, asset A is worth less than asset B. Asset A has an uncertain cash flow which would be discounted at a higher rate, producing a lower present value. Thus, the income yield for A would be higher than B.

Asset B is worth less than assets C, D and E. B has constant income while the others have income growth. All cash flows are certain, so all have the same discount rate. This means that the present value of the income stream from B is lower than that from C, D or E, and so the income yield for B is higher than for C, D or E.

Asset E is worth more than D which is worth more than C. Although each has the same income growth, they have different rates of depreciation (loss of income). The income yield of C is higher than that of D, which is higher than that of E.

Putting all the points in Example 3.6 together gives:

▸ the higher the expected risk, the higher the income yield
▸ the higher the expected income growth, the lower the income yield; and
▸ the higher the expected depreciation, the higher the income yield.

Thus, the market income yield can be used to make deductions about market expectations of risk, growth and depreciation. The formal analysis of income yields is developed in Chapter 5.

3.4 Measurement of risk[4]

While it has only recently been a standard part of the formal analysis of property portfolios, risk is, intuitively, a well-understood concept. Consider two proverbs:

▸ 'A bird in the hand is worth two in the bush' can be regarded as risk aversion and an assessment of the trade-off between risk and return.
▸ 'Don't put all your eggs in one basket' is advice to diversify risk.

Risk is a measure of what is expected to happen, not what is actually happening. Investment decisions require the estimation of the unknown future return. This is known as the *expected return*. It is the best estimate, but there is a range of possible outcomes. Normally, the further from the best guess, the less likely are these possible outcomes. The more likely are extreme outcomes, the riskier is the asset. Figure 3.1 illustrates two possible probability distributions of expected returns. For distribution 1, the best estimate (the most likely outcome) is 10%, but there is a range of possible values. The probability of values in a particular range is given by the area under the curve between the two points.

Figure 3.1▸ Expected return and risk

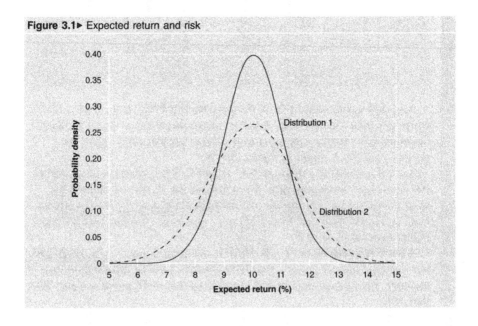

Thus, 9–10% is far more likely than 14–15%. The greater the spread of the distribution, the greater is the area under the curve in the 'tails' and so the greater is the probability of extreme values for return. For distribution 2, extreme values are more likely than under distribution 1.

The spread can be measured by the standard deviation (or its square, the variance) which is the conventional risk measure. This measures the spread of the distribution of expected returns around the best estimate (the mean): the greater the spread, the greater the risk. Strictly, for investment decisions, expected risk is important but this is often proxied by the historical value on the assumption that the spread of historical returns gives a good indication of the spread (the risk) of future returns. It is calculated from a set of observations as:

$$\sigma = \sqrt{\frac{\sum_{i=1}^{n} (x_i - \bar{x})^2}{n}}$$

◄Equation 3.10

where: σ is the standard deviation
$\bar{x}$ (also referred to as $E(x)$) is the average (mean) return;
x_i are the individual observations
n is the number of observations.

Table 3.8 shows the returns for three assets over a ten-year period. A and B have the same mean return but A has a higher risk. B and C have the same risk but B has a higher mean return.

Table 3.8► Return and risk

Year	A	B	C
1	8	7	4
2	12	6	2
3	5	5	1
4	25	6	7
5	12	17	9
6	-9	14	10
7	-1	5	-5
8	2	2	0
9	6	0	10
10	7	5	9
Mean	6.70	6.70	4.70
SD	8.97	5.12	5.12

For most investors, it is important to know the risk that the actual return will be *less than* the expected return; the risk of it being greater is of much less importance. The former is known as downside risk. If the distribution of possible returns is symmetrical about its mean (the expected value) (as in Figure 3.1), then a measure of total risk is also a measure of downside risk. If the distribution is not symmetrical, a separate measure of downside risk is required. There is evidence to suggest that asset returns are not symmetrical but little reason to conclude that this makes a substantial difference to most analyses, and it is conventional to use the standard deviation as a risk measure.

Sources of risk

There are a number of reasons why the actual delivered return could differ from the expected return. Consider Equation 3.3:

$$TR_t = \frac{(CV_t - CV_{t-1}) + NI_t}{CV_{t-1}}$$

The difference between expected and delivered returns could be because of differences between the expected and actual values of any of the right-hand side variables. The capital value is simply the income capitalised at the capitalisation rate (the income yield, k), thus:

$$TR_t = \frac{[(NI_t/k_t) - (NI_{t-1}/k_{t-1})] + NI_t}{NI_{t-1}/k_{t-1}} \qquad \blacktriangleleft \text{Equation 3.11}$$

It was shown in Example 3.6 that the income yield (k) depends on the risk (RP), expected nominal income growth (G) and expected depreciation (d). As will be shown in Chapter 5, it also depends on the nominal risk free rate of return, thus:

$$k = R_N - G + d \qquad \blacktriangleleft \text{Equation 3.12}$$

From Equation 2.1, the nominal discount rate comprises three parts:

$$R_N = RF_R + RP + i_e$$

where: R_N is the required nominal return
$\quad\quad\quad RF_R$ is the risk-free real rate
$\quad\quad\quad RP$ is the risk premium
$\quad\quad\quad i_e$ is expected inflation.

Combining the two equations gives:

$$k = RF_R + RP + i_e - G + d \qquad \blacktriangleleft \text{Equation 3.13}$$

The difference between actual return and expected return can be explained by values of the input variables (or their components) being other than expected:

▶ NI_{t-1} is the income at the start of the period and is known.
▶ NI_t is the income at the end of the period. As property income is fixed for five years, gross income is usually known, although expenses (and so net income) may differ. The exception is where there is a rent review to open market rent which may be different from that expected. The growth in income will depend on supply and demand and on inflation.
▶ RF_R is the real risk-free rate which is taken from the index-linked gilt market. This will vary depending on supply and demand in the market and should affect the capitalisation rates used in the property market.

▶ *RP* is the risk premium which changes according to market circumstances.

▶ i_e is expected inflation which affects both the discount rate and expected nominal income growth.

▶ *G* is the long-term expected nominal income growth rate and expectations may change as expectations of future demand and supply (new development) change and as inflation expectations change.

▶ *d* is expected depreciation which will vary as user requirements and building technology change.

Thus, the actual return may differ from the expected return for a wide variety of reasons linked to the wider economy and to the capital markets. To these must be added valuations. For most properties, the capitalisation rate is not known from a market transaction but from a valuation using a comparison with recent comparable transactions. Errors in the valuation will also cause returns to differ from those expected.

The expected return from a building with a current income of £100k and a capitalisation rate of 10% is estimated on the assumptions that income will remain constant but the capitalisation rate will fall because of an increase in demand and a shortage in supply of quality properties. This produces expected returns as set out in Table 3.9.

Suppose, instead, that net rent falls because of unexpected (and continuing) annual repair costs and that the capitalisation rate rises because a downturn in the economy leads to lower expected income growth. The income returns differ slightly from those expected but the capital value is substantially different. It is affected by both the fall in net rent and by the higher capitalisation rate.

Table 3.9▶ Causes of unexpected return

Variable	Expected	Actual
NI_{t-1}	£100k	£100k
NI_t	£100k	£95k
k_{t-1}	10%	10%
k_t	9%	11%
CV_{t-1}	£100k/0.10 = £1m	£100k/0.10 = £1m
CV_t	£100k/0.09 = £1.111m	£95k/0.11 = £0.863m
Income return	£100k/£1m = 10%	£95k/£1m = 9.5%
Capital return	(£1.111m − £1m)/£1m = 11.1%	(£0.863m − £1m)/£1m = −13.7%
Total return	21.1%	−4.2%

One other aspect of risk which is of particular importance to property investment is that associated with illiquidity. Low liquidity creates two problems: first, it takes longer to realise an asset's market value and, second, there is a risk that the market price will change between the decision to sell and a sale being implemented. Thus, the actual return would differ from that expected at the time of the decision to sell. This adds to the risk premium for property.

Incorporating risk

The conventional way to incorporate risk in an assessment of value is to discount the expected (best estimate) cash flow at a rate which fully reflects the risk. An alternative but little used method is *certainty equivalent cash flows*. This has some value in analysis of development proposals. It may be possible to enter into an agreement with a prospective tenant prior to the development for a rent lower than the expected future market rent. It may also be possible to reach an agreement with a prospective purchaser to sell the completed property at below the expected market value. This has the effect of removing risk and so a risk-free discount rate would be used. A variant on this approach is to use values for the expected cash flows which are regarded as virtually certain. This is an altogether more problematic approach and introduces substantial subjectivity in determining these certainty equivalent cash flows.

In general, it is sensible to undertake *scenario analysis*. This involves using inputs under different future scenarios, most commonly a pessimistic case, a base case and an optimistic case. These scenarios would be based on possible outcomes in the economy and the capital markets and a probability would be ascribed to each. Cash flows and returns would be estimated for the property market under each of these. The expected return and risk can then be calculated using the following formulae:

$$E(x_i) = \sum_{i=1}^{n} p_i x_i \qquad \qquad \blacktriangleleft \textbf{Equation 3.14}$$

$$\sigma(x_i) = \sqrt{\sum_{i=1}^{n} p_i [x_i - E(x_i)]^2} \qquad \qquad \blacktriangleleft \textbf{Equation 3.15}$$

where: $E(x_i)$ is the overall expected return (the mean)
p_i is the probability of scenario i
x_i is the expected return under scenario i
$\sigma(x_i)$ is the overall expected risk.

Example 3.9

Consider the three scenarios for the property market with probabilities and returns as set out in Table 3.10: base, optimistic and pessimistic. Calculate the scenario-weighted expected return and risk.

Table 3.10► Scenario-weighted return and risk

	Scenario			Overall
	Base	Optimistic	Pessimistic	
Probability	60%	30%	10%	100%
Expected return	10%	15%	−10%	$(10 \times 0.6) + (15 \times 0.3)$ $+ (−10 \times 0.1) = 9.5\%$
Expected risk				$[(10\% − 9.5\%)^2 \times 0.6 +$ $(15\% − 9.5\%)^2 \times 0.3 +$ $(−10\% − 9.5\%)^2 \times 0.1]^{1/2}$ $= 6.9\%$

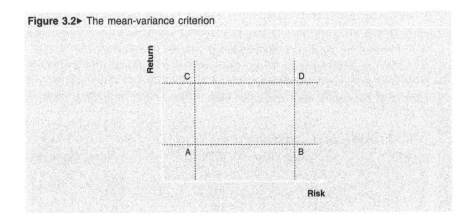

According to financial theory when an investment is considered with an all equity financing (that is, without borrowing), the discount rate will include a risk-free rate and a risk premium. Investment decisions will take place using this discount rate. If debt financing is considered, a premium for financial risk will additionally be considered. Among other things this premium will depend on the degree of gearing (leverage). For shares such a discount rate can be estimated using the Capital Asset Pricing Model (CAPM) (see Chapter 7).

The risk-return trade-off and risk-adjusted returns

One of the fundamental concepts in investment theory is the trade-off between return and risk. Investors are assumed to be risk averse and, other things being equal, require a higher expected return for a higher expected risk. Consider four investments as shown in Figure 3.2. A and B have the same expected return but A has lower risk; similarly for C and D. A and C have the same expected risk but C has a higher expected return; similarly for B and D. A should be chosen instead of B; C instead of D; C instead of A; and D instead of B.

This decision rule is known as the *mean-variance criterion*: the mean (the expectation) is the best estimate of the future return, and its variance (or positive square root, the standard deviation) is a measure of risk. However, this does not assist the choice between A and D, as D has higher expected return but also higher expected risk (this is known as mean-variance efficiency). To decide between these two requires some knowledge of the trade-off an investor makes between return and risk (see Chapter 7).

Risk-adjusted returns are an attempt to provide a single measure which combines risk and return and so allows choice between investments. The most commonly used is simply the expected return divided by the risk. This measures the units of expected return for each unit of risk.

Example 3.10

Consider five investments, as shown in Table 3.11. The risk-adjusted returns show clearly that investment 1 is to be preferred over the others. The problem with this method is that it assumes a linear risk–return trade-off, that is, that investors are prepared to accept the same amount of expected return for each additional unit of risk. In practice, it is likely that investors would require increasing returns for each additional unit of risk. These issues are developed in Chapter 7.

Table 3.11▶ Risk-adjusted returns

Asset	Expected return	Risk	Expected return/risk
1	10	5	2.00
2	10	10	1.00
3	15	15	1.00
4	20	15	1.25
5	25	20	1.25

Combining assets

When assets are combined in a portfolio, the expected return is simply a weighted average of the individual asset's expected returns. The weights are the proportions of these assets held in the portfolio. Thus:

$$E(R_\text{P}) = \sum_{i=1}^{n} w_i E(R_i)$$

◀**Equation 3.16**

where: $E(R_\text{P})$ is the expected return for the portfolio
w_i is the proportion of the portfolio in asset i
$E(R_i)$ is the expected return for asset i.

Example 3.11

Consider the assets shown in Table 3.12.

Table 3.12▶ Portfolio return and risk

	Weight	Expected return	Risk	Correlation
A	0.75	10	10	
B	0.25	5	5	
A/B				0.4

The portfolio expected return is:

$$E_\text{P} = \sum_{i=1}^{n} w_i E_i = 0.75 \times 10 + 0.25 \times 5 = 8.75$$

The portfolio risk is more complex and is given by:

$$\sigma_\text{P} = \sqrt{\sum_{i=1}^{n}\sum_{j=1}^{n} w_i w_j \sigma_i \sigma_j \rho_{ij}}$$

◀**Equation 3.17**

where: σ_p is the portfolio risk

σ_i, σ_j are the individual asset risks (standard deviations)

ρ_{ij} is the correlation coefficient for assets i and j

$\Sigma\Sigma$ is a double summation.[5]

In the above example the portfolio risk is:

$$\sigma_p = \sqrt{\sum_{i=1}^{n}\sum_{j=1}^{n} w_i w_j \sigma_i \sigma_j \rho_{ij}}$$

$$= [(0.75 \times 0.75 \times 10\% \times 10\% \times 1)$$
$$+ (0.25 \times 0.25 \times 5\% \times 5\% \times 1)$$
$$+ (0.75 \times 0.75 \times 10\% \times 10\% \times 0.4)$$
$$+ (0.25 \times 0.25 \times 5\% \times 5\% \times 0.4)]^{1/2}$$

Note that the correlation of an asset with itself is unity.

The portfolio risk depends not only on the weights and individual risks but also on the correlations between assets. These correlations measure the co-movements of asset returns and provide diversification within a portfolio. Unless all of the correlations are equal to unity (the maximum), the risk is always less than the weighted average. Other things being equal, the lower the correlations, the lower the portfolio risk. These issues are considered in detail in Chapter 7.

Other aspects of risk

Two further aspects of risk require mention here. Both are concerned with the range of expected returns in relation to something else, rather than in an absolute sense. These are *tracking error risk* and *liability risk*. These involve, respectively, the risk of losing out to competitors and the risk of being unable to meet liabilities.

In a competitive fund management market, underperforming competitors or the market as a whole is likely to lead to loss of business. This is particularly a problem as fund managers can lose business on the basis of one or two years' bad results rather than a longer term average. Consider two funds: Fund A has a structure identical to its market or competitor benchmark but has high absolute risk (as measured by the standard deviation of expected returns); and Fund B has a lower absolute risk but a structure different from the benchmark. Fund A will always have the same performance as the market. In contrast, Fund B will have a different performance.

Tracking error risk is the standard deviation of expected returns *relative to a benchmark*. Thus, Fund A has no tracking error risk while Fund B has. Other things being equal, the further away from the benchmark, the greater the tracking error risk. As no two properties are identical, so that it is impossible for any fund to hold the same properties as the benchmark, the tracking error has to be modified for application to the property market (see Chapters 4 and 9).

A further aspect relates to fund *liabilities*. For many investors it is essential to be able to match their liabilities with their assets. For example, a pension fund must have the assets to meet its pension payments and an insurance company must have assets to meet insurance claims. Such investors wish to minimise the risk of not being able to meet their liabilities. This adds a new dimension to risk.

One strategy to ensure this is to match the present values of assets and liabilities. One important factor which causes the present value of a cash flow to alter is the discount rate. Different types of cash flows have different sensitivity to changes in the discount rate. Thus, two cash flows may have equal present values for a particular discount rate, but the present values will change in different ways with a change in the discount rate. The sensitivity of the present value to a change in the discount rate can be measured by its *duration*. To minimise this form of risk (not being able to meet liabilities) it is necessary to match the present values and durations of liabilities and assets. Thus, if the discount rate were to change, the present value of both assets and liabilities would change in the same way.

3.5 Summary and conclusions

This chapter has considered the two most important aspects of investment: return and risk. Either can be measured from historical data or can be forecast. In some cases, the historical data is used as a proxy for expected values. Simple measures of return were introduced. These were the income return, the capital return and the total return. For a one-period analysis with income received in arrears, the total return is the internal rate of return.

For more complicated cash flows involving expenditure during the period, the money-weighted rate of return is used. This is an approximation to the IRR if the income is received in arrears. The time-weighted rate of return is suitable for the measurement of fund performance when the fund manager does not have control over new money being allocated to the fund. The different measures of return produce different answers so caution is required when aggregating data and when comparing the performance of different funds.

Two types of yield measure are common across all of the main asset classes. These are the income yield and the equated yield. The former measures initial income return while the latter measures total return. As the former is the ratio of current income to capital value, and as capital value contains market expectations about risk, income growth and depreciation, it is possible to deduce market expectations of these variables from the income yield.

Risk is a measure of what is expected to happen not happening – that is, of the actual return being other than expected. It is measured as the standard deviation of returns, either historical or expected. Often, the historical measure is used as a convenient proxy for a proper expectations measure. The actual return may differ from that expected if any of the inputs into the calculation, or factors driving these, differ from the values expected. For property, such factors are linked to the economy as a whole and to the other capital markets. When forecasting returns, it is often appropriate to consider a number of scenarios for the economic and capital market environment. Valuations and liquidity create an additional element of risk in property investment.

Central to investment theory is the trade-off between return and risk: the higher the expected risk, the higher should be the expected return. It is straightforward to choose between assets with the same expected return but different expected risks, or with the same expected risk but different expected returns. Choosing between two assets, one with low return and risk, the other with high expected return and risk, is more difficult and requires knowledge of the additional return an investor requires for additional risk.

When assets are combined in a portfolio, the portfolio expected return is the weighted average of the individual assets' expected returns, but risk is more complex and involves the correlations between assets. Measures of relative risk are also important: the tracking error measures the volatility of returns relative to a benchmark; and duration measures the sensitivity of asset values to changes in the discount rate. The first measures risk relative to competitors, the second relative to liabilities.

Notes

1. In later chapters it will be shown how the use of valuation estimates rather than actual returns creates problems for this comparison.
2. The term 'total return' is generic and could also apply to an internal rate of return (IRR) calculation. Here it is taken as the MWRR. Most commentators distinguish between the MWRR and discounted cash flow (DCF) analysis. Dubbin and Sayce (1991: 119), however, argue that MWRR is a generic term which can be applied to 'any calculation where income and expenditure are discounted over time to arrive at either an internal rate of return or a net present value'. Throughout this chapter, as is UK market practice, a distinction is drawn between the MWRR and the IRR.
3. The internal rate of return is not without substantial problems. It assumes that all cash flows of whatever magnitude can be re-invested to earn the same (internal) rate of return. In general, great care is required when interpreting IRRs and there can be serious problems if the method is used to rank alternative investment opportunities. Such problems are unlikely to arise for the purposes covered in this book. See Lumby (1988) for a discussion.
4. For a detailed discussion of sources of individual property risk, such as tenant bankruptcy, planning, legal title, see Baum and Crosby (1995, Chapter 2).
5. In a double summation, for each value of i, all the j's are added, then each of these separate sums is added together. Consider the following array of numbers:

		j	
	1	2	3
1	2	3	4
i 2	4	5	6
3	6	7	8

Starting at $i = 1$, add all values of j. Thus, $2 + 3 + 4 = 9$. Repeat for $i = 2$: $4 + 5 + 6 = 15$. Repeat for $i = 3$: $6 + 7 + 8 = 21$. Now add the sums: $9 + 15 + 21 = 45$.

Further reading

Fabozzi, F. J. and Modigliani, F. (1992) *Capital markets: institutions and instruments*, Prentice Hall International Editions, Englewood Cliffs (NJ).

Farrell, J. L. (1997) *Portfolio management: theory and application*, McGraw-Hill International Editions, New York (second edition): Chapters 2, 6 and 15.

Levy, H. and Sarnat, M. (1994) *Capital investment and financial decisions*, Prentice Hall, Englewood Cliffs (NJ) (fifth edition): Chapters 3, 9 and 11.

Ross, S. A., Westerfield, R. W. and Jaffe, J. (1996) *Corporate finance*, Irwin, Chicago (fourth edition): Chapters 4, 6, 9 and 10.

Sharpe, W. F., Alexander, G. J. and Bailey, J. V. (1995) *Investments*, Prentice Hall International, Inc., Englewood Cliffs (NJ) (fifth edition): Chapters 18 and 25.

Young, M. S., Geltner, D. M., McIntosh, W. and Poutasse, D. M. (1995) Defining commercial property income and appreciation returns for comparability to stock market-based measures, *Real Estate Finance*, 12(2), 19–31.

Chapter 4

Property indices

Introduction

In the previous chapter the focus was mainly on individual assets. For several reasons, however, it is necessary to combine information on a population, or a sample, of assets in order to have an aggregate picture. The evolution in time of asset prices and returns is monitored by indices. Indices are widely used as measures of economic activity. Inflation, for instance, can be measured by the Retail Price Index (RPI) in the UK and the Consumers' Price Index (CPI) in the USA. Several indices also exist for shares, bonds and, to a lesser extent, property.

Indices for the main asset classes (shares, bonds and property) are needed for a variety of reasons. They make it possible, for example, to compute holding period returns which enable, in turn, the calculation of the average return and standard deviation for each of the asset classes. Moreover, the correlation coefficient between each pair of assets can be calculated. With data on the returns and standard deviations for each asset class, and the correlation coefficients between each pair of assets, it is possible to develop portfolio strategies (see Chapters 7 and 9) and to gain knowledge as to the optimal weight which should be allocated to each class of asset in an investment portfolio (see Chapter 10). The returns data also make it possible to investigate the inflation-hedging effectiveness of the various asset categories. Finally, indices are used as benchmarks in performance measurement analyses and in portfolio strategy (see Chapter 9).

Several issues have to be addressed when indices are constructed. First, a decision has to be made whether the index will encompass the whole market or whether a sample only will be considered. For the UK share market, for instance, the FTSE 100 comprises the 100 largest shares by market capitalisation, while the FTSE Actuaries All-Share index contains all shares quoted on the main stock market.[1] As always, the decision on sample size is a trade-off between accuracy on one hand and cost and availability of data on the other.

A further issue in index construction is whether an income return index, a price index or a total return index is required. An income return index will only consider the income on investments, a price index the price change, while a total return index will consider both the income and the price components. Another issue concerns the weighting of assets in the index. The two methods which have been traditionally used are the market capitalisation weighting method and the equally weighted method. With the former method, the weight for each asset is obtained by dividing the market value of that asset by the total market values, whereas with the latter method each asset is awarded the same weight.

For shares and bonds, information concerning income and price is readily available. These securities are usually traded on exchanges, often on a continuous basis. This information is available in some daily newspapers, in on-line systems and in financial databases. Owing to the abundance of data on securities prices, indices exist in almost every country. In fact, in most countries several stock market indices are available. If price changes only are considered, the indices are price indices, while they are total return or performance indices when both price changes and income are considered. The main focus in this chapter is on property price indices, but rent indices are also available. These are discussed in Chapter 6.

The construction of property indices raises issues not encountered in the securities markets. These are mainly the result of the characteristics of property and the property market considered in Chapter 2. The first important problem is defining the property market. Conventionally, in the UK, residential property is treated separately as it is mainly owner-occupied or publicly rented. Other types or sectors of property, such as hotels, pubs, agricultural land and forestry land, feature in the portfolios of some funds but these markets are dominated by specialist investors or owner-occupiers rather than the institutions. The major investment portfolios in the UK consist predominantly of 'prime' offices, shops and industrials. Accordingly, the main UK property indices comprise such property types to the virtual or total exclusion of others. Other countries have different property investment markets. For instance, in Switzerland where the percentage of households which own their property is low, institutions play an important role in the residential sector. Swiss property indices, therefore, mainly comprise residential property (apartment buildings).

Another problem when constructing property indices lies in sample selection. The heterogeneity of property investments (see Chapter 2) means that index construction is complex. Any two indices based on different samples are likely to produce different index values as a result of the specific characteristics of properties. A third and related problem is the effect of the size of the sample. As the sample size increases, the risk of the individual assets will be diversified and the risk of the index will approach the market risk, thus providing for a better picture of the market (see also Chapter 7). Finally, the lack of a central trading market with price information for property, and infrequent trading of individual properties also creates a problem. Each property is bought and sold infrequently at different times. Thus, it is in most cases impossible to compare the prices of a sample of properties on either a regular or a frequent basis. In some cases it is also impossible to obtain information on transaction prices.

The above issues make the construction of a price or return index in any property market difficult and, in the commercial property market, more problematic. However, the problems are different in the residential market where the population is large and properties usually can be subdivided into groups with a reasonable degree of similarity within a group. An example of such a group is terraced houses in a particular geographical area. In the commercial property market, the insufficient amount of information available on transaction prices has led to the construction of indices based on regular valuation of a sample of properties. These are considered in Section 4.2. Methods based on transaction data and which are mostly utilised in the residential sector are presented in Section 4.3. These methods include the computation of averages of transaction prices, the hedonic method and the repeat sales method. As an alternative to focusing on direct property investments, it is also possible to consider property company shares. These vehicles are considered in Chapter 11, but the advantages and disadvantages of using such data are considered in Section 4.4. Section 4.5 contains a summary and conclusions.

4.2 Appraisal-based indices

General approach

Appraisal-based indices are well suited for markets in which price information is scarce and, therefore, exist mostly for commercial markets.[2] As it is impracticable to consider the whole market, appraisal-based indices rely on a sample of properties. The starting point of this method is thus to collect data for a sample of properties which are as representative as possible of the property market. An index can be constructed for a national market and for particular property types or for a region only.

As the properties in the sample are not sold during each period, the value of the properties included in the sample has to be estimated on a regular basis. If a quarterly index is required, the properties will be appraised every quarter. As can be imagined, estimating the value of properties on a regular basis can be a large task. From a practical perspective, this problem can be solved by selecting, for the index, properties that belong to institutions which are for other reasons legally required to estimate the value of their properties on a regular basis. In the UK, for instance, the Investment Property Databank (IPD) annual index contains information on the portfolios of most major institutional investors. These data are collected for annual performance measurement purposes. The sample on which the IPD monthly index is based is much smaller. It is based on properties which belong to unitised funds and which require monthly valuations. Other indices exist, the most important of which is the Jones Lang Wootton (JLW) quarterly index which comprises buildings in funds under JLW management.

As mentioned, appraisal-based indices rely on valuations of the properties included in the sample. Consequently, some knowledge of valuation methods and an understanding of the problem of valuation smoothing are needed. A brief overview of valuation methods is provided below.

Overview of valuation methods

The purpose of most valuations is to estimate the market value of a property at a given point in time. Although valuations can also aim at estimating the insurance value or the tax value of a property, in the case of appraisal-based index construction the aim is clearly to estimate the market value of the properties included in the sample at each point of time. The following definition of market value for use in appraisal reporting has been agreed by federal financial institutions in the USA (see Boykin and Ring, 1993: 11–13)[3]:

'The most probable price which a property should bring in a competitive and open market under all conditions requisite to a fair sale, the buyer and seller acting prudently and knowledgeably, and assuming the price is not affected by undue stimulus. Implicit in this definition is the consummation of a sale as of a specified date and the passing of title from seller to buyer under conditions whereby:
1. buyer and seller are typically motivated;
2. both parties are well informed or well advised, and acting in what they consider their best interests;
3. a reasonable time is allowed for exposure in the open market;
4. payment is made in terms of cash in United States dollars or in terms of financial arrangements comparable thereto; and
5. the price represents the normal consideration for the property sold unaffected by special or creative financing or sales concessions granted by anyone associated with the sale.'

A similar definition of an open market valuation for the UK can be found in the Royal Institution of Chartered Surveyors (1995) 'Red Book' (Practice Statement 4.2)[4]:

'An opinion of the best price at which the sale of an interest in property would have been completed unconditionally for cash consideration on the date of valuation, assuming:
1. a willing seller;
2. that, prior to the date of valuation, there had been a reasonable period (having regard to the nature of the property and the state of the market) for the proper marketing of the interest, for the agreement of the price and terms and for completion of the sale;
3. that the state of the market level of values and other circumstances were, on any earlier assumed date of exchange of contracts, the same as on the date of valuation;
4. that no account is taken of any additional bid by a prospective purchaser with a special interest; and
5. that both parties to the transaction had acted knowledgeably, prudently and without compulsion.'

Three methods are generally used to estimate the value of properties. These methods include the sales comparison approach, the depreciated cost approach and the income approach.

Sales comparison approach

The sales comparison approach estimates value by comparing the subject property to comparable properties that have sold recently. The sales price of the comparable properties is adjusted for differences between the comparable property and the subject property. The purpose of the adjustments is to estimate the price at which the comparable property would have sold if it were the same as the subject property. For example, if the comparable property had a better location than the subject property, then, other things being equal, it should have a higher value than the subject property. Thus, the price of the comparable property would have to be adjusted downwards to provide an indication of the value of the subject property. Common adjustments include differences in the physical characteristics of the properties, differences in location, differences in lease contracts, in the quality of the tenants and in quantum, and any change in market conditions from the time that the comparable property sold.

The price of each of the comparable properties is often divided by an appropriate unit of comparison as a first step in the adjustment process. The units used to compare prices differ for different property types. The unit of comparison which is typically used is the price per square foot for flats, houses, shops and industrial properties, the price per cubic foot for warehouse space, the price per seat for cinemas, the price per bed for hotels, the price per parking space for car parks and so on.[5]

Depreciated cost approach

This method is based on the premise that the value of the property is equal to the sum of the land value plus the depreciated replacement cost of the building. The economic rationale for the depreciated cost approach is that a rational investor would not pay more than the cost of replacing the building with one that is equally productive on a comparable site. Thus, the value estimate (VE) should amount to the land value (LV) plus the replacement cost (new) of the building (RC) minus depreciation (d):

$$VE = LV + RC - d$$
◀Equation 4.1

The value of the land is usually estimated by means of the sales comparison approach. The replacement cost is typically calculated by multiplying the number of square feet of the building by the replacement cost per square foot. Depreciation is the loss of value of the subject property when compared to a new property and is discussed in Chapter 8.

Income approach

The idea of the income approach is to determine what the typical investor would be willing to pay for the stream of income that is expected from the property. The valuer must estimate the market rent for comparable space as well as the impact of lease terms, and the typical expenses for the given type of property.

This method would normally require that all cash flows for the economic life of a property be estimated and then discounted at the appropriate rate (discounted cash flow analysis). In practice, simplified methods are generally used. One of these methods is to divide the first year net operating income (NOI) from the property by the capitalisation rate. As will be seen in Chapter 5, the capitalisation rate comprises the required return (R) minus the growth rate (G) plus the depreciation rate (d):

$$V = \frac{NOI}{R - G + d}$$
◀Equation 4.2

Depreciation is often deducted from the growth rate, and the growth rate becomes a net rate. This is a variant of Gordon's growth model which is often used for valuing shares.

A hybrid method involves deriving the capitalisation rate from comparable sales, in which case the capitalisation rate is found by dividing the *NOI* of comparable properties by their sales prices.

Reconciliation of estimates of value

It is unlikely that the three approaches used to estimate the value of a property will give the same answer. The valuer will thus have to use his/her judgement to derive one single figure that will be his/her estimate of value for a given property. As stated by Boykin and Ring (1993: 434), 'reconciliation is the careful weighing of the initial value results on the basis of accuracy and completeness of data and in light of market conditions that prevail on the date of the appraisal'.

The reconciliation method differs quite substantially from one country to another. In the USA, all three methods are used in most cases. Boykin and Ring (1993: 434) mention that:

> 'although the importance of each of the three approaches to value may vary, depending on the kind of property and the purpose that the appraisal is to serve, nevertheless, it is important to consider each approach to value as a separate entity under the appraisal process and to reach independent value conclusions in relation to replacement costs less accrued depreciation, market sales of comparable properties, and capitalization of net operating income derived from property operation under typical ownership and management.'

Depending on the availability of data and the type of property being valued, one approach may be more reliable than another. When significant differences exist, the procedures should be checked for accuracy. If the differences remain after rechecking, then the valuer should use his/her experience, judgement and skill.

In France, only the comparable sales and the income methods are used. In Switzerland, the depreciated cost approach and the income method are used. The final estimate of value is obtained by computing a weighted average of the two value estimates. The weights used are, in most cases, one-third for the depreciated cost approach and two-thirds for the income approach. In the UK, a hybrid method is predominantly used. It divides a known rent by a capitalisation rate obtained by market comparison.

Indices available

Appraisal-based indices exist in several countries. In the USA, the NCREIF Property Index (NPI, formerly the Russell–NCREIF index) is the most widely used index. It has existed since 1978 and tracks institutional-grade commercial property returns. It is based on the appraised values of ungeared properties held for institutional investors in the portfolios of the member firms of the National Council of Real Estate Investment Fiduciaries (NCREIF). As of the second quarter of 1998, the index included 2438 properties appraised at $65.7bn. Sub-indices also exist for apartments (the sample includes 532 apartments valued at $10.5bn), industrial properties (722 properties valued at $10.6bn), offices (646 properties valued at $25.9bn) and retail (511 properties valued at $18.0bn). Sub-indices by geographical region are also available. The indices are computed on a quarterly basis.[6]

Several appraisal-based indices are available in the UK (see Morrell, 1991 for a survey). The most commonly used indices are the IPD (Investment Property Databank) index, the JLW (Jones Lang Wootton) index and the ICHP (Investors Chronicle Hillier Parker) index.[7] The IPD annual index is the main UK property index. It has by far the largest sample and comprises over 12,000 properties, valued at over £51bn at the end of 1996, and making up 42% of the total investment (as opposed to owner-occupied) market. Other UK property indices are much smaller. Sectoral (retail, office and industrial) and a minimum of 12 geographical indices are also computed by IPD.

An appraisal-based index, called the IPD BD2i (Base de Données des Investissements Immobiliers) index, has recently been created in France (but has been backdated to 1984). As of the end of 1997, the sample included approximately 3000 buildings with 12 million square metres and a total value of Ffr140bn. The sample is representative of the holdings of French insurance companies. The main categories of properties in the sample are residential properties (44.6% of the total value of the buildings in the sample), offices (36.5%) and shops (6.3%). Ninety per cent of the buildings are located in the greater Paris region (Île-de-France). The frequency of the index is annual.

In the Netherlands, the ROZ/IPD Netherlands index was created in 1995. As of the end of 1997, 5350 buildings were in that index. The total value of the buildings amounted to NLG51.1bn. Sub-indices for retail, offices, residential properties, industrial, mixed use and other properties are available. The frequency of the indices is annual. The IPD also computes indices for Germany, Ireland, South Africa and Sweden. In Canada, the Russell Canadian Property Index (RCPI) has existed since the beginning of 1985.[8] As of the end of 1997, the index included 897 buildings, with an estimated market value of Canadian $15.5bn. The index is computed on a quarterly basis and sub-indices by province/region and property type are available.

Appraisal-based indices also exist in Australia (Property Council of Australia, PCA, formerly BOMA), Hong Kong (JLW Hong Kong Index), New Zealand (BOMA New Zealand) and South Africa (Richard Ellis).[9]

The smoothing issue

Appraisal-based series are smoothed which means they understate the variability of returns in the property market (see also Chapter 10). Smoothing enters index returns from two sources. First, infrequent transactions leave appraisers with little information to work with in determining market value at specific times. This leads appraisers to combine indications of value from the most recent comparable sale with past appraised values to arrive at the value that is actually reported for a given building each period. Second, in addition to any smoothing introduced at the disaggregate level by the appraisal process, aggregation of property values within an index causes additional smoothing. If property values are appraised at different points in time throughout each calendar quarter, yet all these valuations are, in effect, averaged together in the index to produce the index value attributed to that quarter, then the index value will be a moving average of spot values (see Geltner, 1993a).

Fisher *et al.* (1994) argue that:

'we might expect that over sufficiently long periods of time, the average appreciation return displayed by the Russell–NCREIF index would be a relatively unbiased estimate of the true appreciation in the typical property (of the type represented in the index). However, the variance of short-interval returns across time, and the covariance of short-interval returns with contemporaneous returns on other assets, will be biased toward zero, and the index will tend to lag underlying property market value changes, due to these sources of smoothing. Furthermore, the smoothing phenomena described above would be expected to add significant positive autocorrelation (that is, apparent inertia or self-predictability) into the Russell–NCREIF return series.'

Appraisal-based property indices thus represent smoothed versions of a true underlying price series. Blundell and Ward (1987) and Ross and Zisler (1991) suggest methods for correcting for smoothing. These methods transform property returns derived from valuations to market price proxies. Unsmoothing is based on the assumption that the appraised value is a weighted average of the true series at times t and $t - 1$ and that the weights are constant:

$$r_t^u = \frac{r_t^* - (1 - a)r_{t-1}^*}{a}$$

◄Equation 4.3

where: r_t^u is the unobservable underlying property markets return series at time t
r_t^* is the observed appraisal-based index return in year t
a is a smoothing parameter.

Thus, the true series can be extracted from consideration of the coefficient of first-order serial correlation (see Ball *et al.*, 1998, for a fuller discussion).

As will be seen in Chapter 10, the risk of property as measured by the returns' standard deviation is greater when de-smoothed property indices are used. MacGregor and Nanthakumaran (1992) use the method of Blundell and Ward to de-smooth the JLW quarterly return series. The standard deviation of the real returns on the transformed series is nearly double that of the smoothed series. A similar analysis of the nominal quarterly series suggests an increase by a factor of 2.86. Ross and Zisler (1991) suggest a factor of 3–5 for US data using quarterly nominal returns on the Russell–NCREIF index. Brown's (1991) analysis of IPD monthly nominal returns (1987–90) shows a factor of 3.44. As pointed out by MacGregor and Nanthakumaran (1992), other data series are almost certain to have different multipliers (see Chapter 10 for a fuller discussion).

The de-smoothing method proposed by Blundell and Ward (1987) assumes that real estate markets are informationally efficient. Geltner (1993b) develops another approach to de-smoothing the Russell–NCREIF index, which avoids making the efficient market assumption. By positing a plausible model of appraisal smoothing at the disaggregate level, and incorporating the effect of temporal aggregation and seasonality at the aggregate index construction level, a quantitative structural model of smoothing in the Russell–NCREIF index is developed without resorting to the efficient market assumption.

Fisher *et al.* (1994, 157) report a standard deviation of 5.20% for the Russell–NCREIF index, a standard deviation of 8.62% for the de-smoothed series using a technique similar to that developed by Blundell and Ward (1987) and a standard deviation of 8.19% for the de-smoothed series using the Geltner (1993b) technique. These figures are for annual real returns for the period 1979–92. For the UK, a similar approach is used by Barkham and Geltner (1994) to de-smooth the JLW index of capital values.

4.3 Indices based on transaction prices

When more information is available on property transaction prices, as is usually the case in the residential market, property indices based on transaction prices can be constructed. A simple way of constructing property indices based on transaction prices is to compute *averages of transaction prices*. With such indices, the characteristics of the properties and the characteristics of the neighbourhoods in which the properties are located are not taken into consideration. A preferable approach is to construct quality-adjusted or quality-constant property indices. This can be done by using the *hedonic method* which takes into account the characteristics of the properties, or the *repeat sales method* which only considers properties that have sold at least twice over a given time period. Each of these three methods is reviewed below.

Averages of transaction prices

In order to filter out some of the heterogeneity of property assets, the starting point of this method is to define a homogeneous segment of the real estate market. One property type in one town or region is usually considered. In France, for example, the Notaires/Insee index is widely used. This index covers flats in Paris which are at least five years old. The segment of the market is, thus, relatively homogeneous, at least far less heterogeneous than if the whole French property market were considered.

Once the segment has been defined, information has to be collected on transaction prices of the properties which have been sold during each time increment. If an annual index is required, then the prices are collected for each year. For quarterly indices, data are collected on a quarterly basis and so on. In most cases, the transaction prices are divided by the square footage or the number of square metres in the property, so data on the size of the properties are also required. The average price (or price per square foot) for each time increment (year, quarter, month) can then easily be computed. The index is then simply constructed on the basis of the average price (or price per square foot) for each time increment.[10]

As stated above, the Notaires/Insee index of flats in Paris that are at least five years old is widely used. Indices have also been constructed on an ad hoc basis, for example, for apartment buildings in Geneva (see Bender *et al.*, 1994). Such indices exist also in several countries for vacant land.

The Halifax Building Society publishes quarterly averages of house prices in the UK. These averages are available for 12 regions (Scotland, Northern Ireland, North West, West Midlands, Wales, South West, North, Yorkshire and Humberside, East Midlands, East Anglia, Greater London and South East), for various types of houses (terraced houses, semi-detached houses, detached houses, bungalows and flats and maisonnettes) and for various construction periods (pre-1919, 1919–45, 1946–60, post-1946 but not new, and new). For semi-detached houses, town and county data are also available.

This method has two main advantages. First, the data for each transaction are limited to the sales price and possibly the size of the building (or the size of the land if an index of vacant land is constructed), which makes the data-collection process far less onerous than if several characteristics of the properties, such as the quality of the construction, the quality of the neighbourhood and so on, had also to be collected. Second, such indices are simple to construct and do not require special econometric methods.

This method, however, has a severe drawback which stems from the heterogeneity of real estate assets. Although the market is limited to a homogeneous part of the market (for example, flats in Paris which are at least five years old for the Notaires/Insee index), several sources of heterogeneity remain. These sources include the condition of the building, the quality of construction, the quality of the neighbourhood, the quality of the location within the neighbourhood and so on. It might well be that transactions during one period of time are of rather 'good' quality, whereas the transactions in the next period are of much poorer quality. As stated in the Halifax bulletins, 'prices shown are the arithmetic average prices of houses on which an offer of mortgage has been granted. These prices are not standardised and therefore can be affected by changes in the sample from quarter to quarter.'

When the real estate market is bearish, for example, few transactions occur and may involve buildings which are in poor condition. If this is true, a decline in the value of the index from period $t-1$ to period t could reflect a decline in real estate prices and/or the fact that the buildings which are sold in period t are of poorer quality than the buildings sold during period $t-1$. The reverse also holds: the index can rise for one or the other reason or for both reasons together. Thus, it is impossible, on the basis of such indices, to ascertain the true evolution of prices in real estate markets. As stated above, this is a severe drawback and it explains why such indices are seldom used in the literature in order to gain knowledge on the return and risk of property investments (see Chapter 10 for a review of this literature).[11]

The hedonic method

General approach
Both the hedonic method and the repeat sales method have been devised in order to control for the heterogeneity of assets and, thus, make it possible to construct quality-constant indices. Although the hedonic methodology was first developed to track automobile prices (Court, 1939; Griliches, 1961), the model is very well suited for property. This approach recognises that a property is a composite product: while the attributes are not sold separately, regressing the attributes on the sales price of the composite product yields the marginal contribution of each attribute to the sales price. The theoretical foundation of this method was developed by Lancaster (1966) and Rosen (1974).

Both physical characteristics (of the land and the building) and locational characteristics have to be considered in hedonic models. Physical characteristics include the size of the land, the size of the building (area or number of rooms), the year of construction, the quality of construction, the condition of the building and the number of bathrooms. Locational variables are concerned with accessibility, the quality of the neighbourhood and the quality of location within the neighbourhood. For apartment buildings in Geneva, for instance, Büchel and Hoesli (1995) found that the floor area, the age of the property, the average number of bathrooms in a flat, the presence of a lift, the condition of the building, the distance from the city centre, the quality of location and whether or not the buildings were subsidised had to be taken into consideration in a hedonic model.

The location variables are qualitative, meaning that a judgement has to be made. A scale is used (1 to 3, or 1 to 4, or sometimes even 1 to 10) and the quality of location is measured on that scale. In order to reduce part of the judgement, quantitative variables are being used more and more in order to measure the quality of location instead of qualitative judgements. Examples of quantitative variables which can be used to measure the quality of a neighbourhood for residential properties are the noise level, distance to the city centre, distance from the station, distance to shops, distance to schools, and distance from a park. If a Geographic Information System (GIS) exists, such variables can be measured by means of that system.[12]

The hedonic method has two very important uses in property. First, it can be used to estimate the value of a property on the basis of its characteristics. Second, it can be used to construct property price indices. When the hedonic method is used for valuation purposes, the characteristics of a property are valued using the regression coefficients (the price of one unit of characteristic) and the price of a property can be estimated by summing the price of the characteristics. Such models are being used on a wide basis in Switzerland to estimate the value of residential properties and the value of property portfolios of banks and institutions.

A simple hedonic model for flats yields the following results:

$$V = 5000 + 80S + 6000L$$

where: V is the value of the flat
S is the size of the flat (in square feet)
L is the quality of location (on a scale from 1 to 3).

The value of a flat with 2000 square feet in a very good area (quality of location = 3) would thus equal:

$$V = 5000 + (80 \times 2000) + (6000 \times 3) = £183,000$$

and the value of a flat with 500 square feet in a bad area (quality of location = 1) would equal:

$$V = 5000 + (80 \times 500) + 6000 \times 1 = £51,000.$$

For price indices, the hedonic regression model can be applied in two different ways (Crone and Voith, 1992). First, the price coefficient of each structural and locational characteristic is estimated by means of a cross-section regression. Such a regression is re-estimated for each time period t. Then, a property with standard characteristics (for example, the average characteristics of the properties included in the sample) is valued for each time period and an index is constructed.

The second alternative way of applying the hedonic model is to estimate a panel regression (combining the cross-section and time series data), including dummy variables for the various time periods. For each property, the dummy variable for a given period will be equal to 1 if the property sold during that period and equal to 0 otherwise. The price changes from the base year can be inferred from the coefficients on the time dummy variables.

Assume that a hedonic price index for the period 1989–99 is constructed. Annual data are used. Dummy variables are needed for each period (in this case, each year) except for the base year (1989 in this case). Thus, there will be one dummy variable for 1990 (T_{90}), one for 1991 (T_{91}), and so on until 1999 (T_{99}). The values of the time dummy variables for a property which was sold in 1992 and another property which was sold in 1997 are shown in Table 4.1.

Table 4.1▶ Time dummy variables for constructing a hedonic index

	T_{90}	T_{91}	T_{92}	T_{93}	T_{94}	T_{95}	T_{96}	T_{97}	T_{98}	T_{99}
Any property sold in 1992	0	0	1	0	0	0	0	0	0	0
Any property sold in 1997	0	0	0	0	0	0	0	1	0	0

Hedonic models are mostly used in order to construct price indices, but some total return indices also exist. This is the case, for example, in Switzerland for apartment buildings. This is because the percentage of owner-occupants in Switzerland is quite low compared to other countries. Thus, data on rents are more readily available. Swiss institutions are large players in the Swiss residential market and seek total return indices for apartment buildings.

Hedonic indices
In addition to constructing indices of averages of transaction prices (see above), the Halifax Building Society computes a quarterly index of UK house prices using the hedonic method. An index for each of the 12 regions is also available. The quarterly index is available for all houses, new houses and existing houses. It also exists for first-time buyers and for former owner-occupiers. National monthly indices are also calculated for all houses, new houses and existing houses. First-time buyers and former owner-occupiers indices are also available. All indices are set at 100 as of 1983. However, these indices have limited hedonic variables.

The hedonic method has also been widely used in the USA to construct price indices. Thibodeau (1992), for instance, reports house price indices for 60 US metropolitan areas surveyed between 1974 and 1983. Other studies for the USA include Hoag (1980), Miles et al. (1990), Webb et al. (1992), Fisher et al. (1994), Meese and Wallace (1997) and Kiel and Zabel (1997).

Figure 4.1▶ Price indices for Swiss single-family houses and flats (Q3/96–Q4/98)

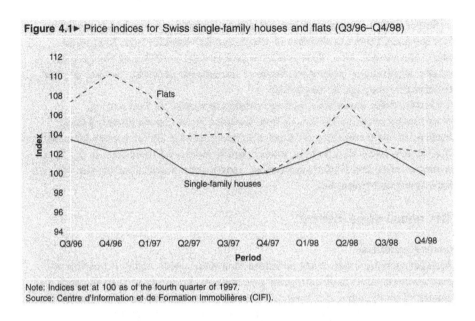

Note: Indices set at 100 as of the fourth quarter of 1997.
Source: Centre d'Information et de Formation Immobilières (CIFI).

The study by Fisher *et al.* (1994) is of particular interest in that these authors compare four types of indices for the USA: the unsmoothed Russell–NCREIF index, an index of the average of the net operating income on properties divided by an average capitalisation rate, a hedonic index and an index based on ungearing REIT share prices (indirect property). Their results show that the REIT index leads the other indices in time but displays greater short-run volatility (see also Chapter 10), while the two transaction-based indices lag behind the other series. The study by Meese and Wallace (1997) suggests that the hedonic method is better suited than the repeat sales method (see below) for the construction of property price indices.

The hedonic method is also used in several other countries. Lin (1993), for example, uses it for Taiwan. For Switzerland, Hoesli, Giaccotto and Favarger (1997) use the hedonic method to construct price indices of residential buildings and vacant land in Geneva. The hedonic method is currently being used for the Swiss residential market by the Centre d'Information et de Formation Immobilières (CIFI), a Zurich-based company. The CIFI constructs hedonic price and total return indices, but also valuation models for properties and portfolios of properties. Figure 4.1 contains the price indices for single-family houses and flats for the period from the third quarter of 1996 to the fourth quarter of 1998 (both indices are set at 100 as of the fourth quarter of 1997).

Advantages and disadvantages of the hedonic method

The hedonic method is very attractive as it relies on transaction data (rather than appraisals) and corrects for the effects of the heterogeneity of assets by taking the characteristics of the assets into account. The resulting indices should, therefore, monitor price changes in a reliable manner. A further advantage of the hedonic method is that it permits valuation of properties based on transaction prices and not on judgements by valuers.

Shiller (1993), however, points out some problems with the hedonic method. For instance, there is a decision of which quality variables to include in the regression model. Also, there is often a lack of data available on the various quality attributes of properties. Some of the hedonic variables that are relevant to valuation may not be observable.

Despite these limitations, hedonic models represent the best way of constructing property indices, as demonstrated by the development of these models for Switzerland by the Centre d'Information et de Formation Immobilières (CIFI). It is likely that hedonic models will be more and more used in the future to construct price and total return indices, mostly for residential properties but also for commercial properties.

The repeat sales method

General approach

Another technique that relies on transaction data, which makes it possible to construct constant quality property price indices, is the repeat sales method. As suggested by its name, this method relies on transaction prices of properties that have sold more than once during a given time period. This approach was first suggested by Bailey et al. (1963), and was further developed by Palmquist (1982) and Case and Shiller (1989).

Intuitively, the repeat sales method is appealing as the effects of the heterogeneity of assets should be greatly diminished as each price change is measured on the same property. The price changes are obtained by regressing the logarithm of the ratio of second sales price to first sales price on a set of dummy variables that are equal to −1 for the period of first sale (unless the first sale occurred in the base period) and to 1 for the period of second sale. For all other years, the value of the time dummy variables will be 0. If a property which sold three times during a given time period is included in the sample, there will be two repeat sales: one between the first and second sales and one between the second and third sales. The coefficients on the dummy variables indicate the price change from the base year.

A repeat sales index is to be constructed for the period 1989–99. One property (Property 1) was sold in 1990 and 1997, while the other (Property 2) was sold in 1989, 1991 and 1999. Calculate the values of the time dummy variables.

As the second property sold three times, there are two repeat sales for that property. As with the hedonic method, there is no time dummy variable for the base year (in this case 1989). The values of the time dummy variables are given in Table 4.2.

Table 4.2► Time dummy variables for constructing a repeat sales index

	T_{90}	T_{91}	T_{92}	T_{93}	T_{94}	T_{95}	T_{96}	T_{97}	T_{98}	T_{99}
Property 1	−1	0	0	0	0	0	0	1	0	0
Property 2 (first repeat sale)	0	1	0	0	0	0	0	0	0	0
Property 2 (second repeat sale)	0	−1	0	0	0	0	0	0	0	1

Several US authors have made use of the repeat sales method to construct real estate price indices for residential properties (see, for example, Bailey et al., 1963; Case and Shiller, 1989; Abraham and Schauman, 1991; Dombrow et al., 1997; Follain and Calhoun, 1997; and Meese and Wallace, 1997). The study by Case and Shiller, for instance, encompasses nearly 40,000 repeat sales (8,945 in Atlanta, 15,930 in Chicago, 6,669 in Dallas and 8,066 in San Francisco). Recently, this method has been applied to flats in Geneva by Hoesli, Giaccotto and Favarger (1997). The repeat sales method has also been used for commercial real estate in Florida (Gatzlaff and Geltner, 1998).

Advantages and disadvantages of the repeat sales method

The main advantage of the repeat sales method is that the characteristics of the properties are not needed. Only the transaction price and the date of sale are needed for each property. This method, however, has severe drawbacks. Some properties may sell repeatedly because they are speculative or because they have a flaw. Clapp et al. (1991), for example, report that properties in the Hartford, Connecticut, area that sell twice have average sales prices about 15% lower than those that sell only once. Properties that sell three or more times have sales prices about 7% lower than those that sell only twice.

Using the Case and Shiller (1989) data, Clapp et al. (1991) also compare properties that sell three or more times to those that sell twice. They find a 5% difference in Chicago and Dallas, a 20% difference in Atlanta and a 50% difference in Oakland. All of these differences indicate lower sales prices for properties that sell more frequently.

Further, only a few buildings sell more than once, making it difficult to have sufficient data to use this technique. Abraham and Schauman (1991) found 2.5% of repeats over a 19-year interval. Case and Shiller (1989) found that 4.1% of transactions were repeats over a 16-year interval. Mark and Goldberg (1984), however, found 40% of repeats over a 22-year period. Finally, the assumption that quality of the assets' attributes are unchanged over time may not hold. Other attributes of the properties may also have changed with time. This is the case, for example, for the condition of the building, the quality of the neighbourhood or the quality of the location within the neighbourhood. All of these may have changed in a positive or a negative way. The condition of the building may have deteriorated due to wear and tear and obsolescence (see Chapter 8) or may have been improved by renovation of the building. The quality of a neighbourhood can improve, for instance, if it becomes fashionable to live in that area and there is a net gain of people with high incomes. The quality of the location within a neighbourhood can improve, for example, if a noisy road gets diverted due to the construction of another road which does not affect that part of the neighbourhood. Similar examples for adverse impacts on the quality of a neighbourhood or the quality of the location within a neighbourhood can easily be found.

To deal with these criticisms of the repeat sales method, recent work has focused on combining hedonic models and repeat sales models to improve estimation precision. Case and Quigley (1991), for instance, devise a technique which makes it possible to use all transactions data and produces efficient house price indices for properties that have sold at least once. Other authors who combine both methods include Clapp and Giaccotto (1992), Hill *et al.* (1997), Quigley (1995), Eichholtz (1997a) and Meese and Wallace (1997). Knight *et al.* (1995), for instance, combine the hedonic and repeat sales models in a way that allows parameter coefficients to vary from period to period.

4.4 Indices of property company shares

Given the difficulties of constructing indices for direct property investments due to the scarcity of data, it may seem appealing to use data for indirect property investments. As will be discussed in Chapter 11, an investment in indirect equity property can be realised through the purchase of shares of companies which invest in property or trade or develop property. Thus, an interest in property is not acquired directly by the purchase of buildings, but is obtained indirectly by the purchase of shares of companies which own property. Provided indirect property investments behave in a similar fashion to direct investments, indices of indirect property can be used as benchmarks for direct property. This debate has been the focus of much research. In any case, indices of indirect property are very useful as benchmarks for investors who own such shares.

The design of real estate securities varies quite substantially from one country to another (see Chapter 11). In the USA, real estate securitisation is achieved through Real Estate Investment Trusts (REITs). A REIT is a specialised form of business trust that owns property or investments in property. Unlike US REITs, UK property companies have no particular legal form or specific tax advantage, but are ordinary public companies. Some companies invest in income-producing properties, while others develop or trade properties. Property companies in other countries exhibit a wide diversity according to the following dimensions:

▶ type of activity (investment companies or developers/traders)
▶ type of investment (residential, commercial property or both)
▶ region of investment (domestic or international)
▶ tax status (tax exempt or not)
▶ gearing (leverage).

Indices of indirect property are widely available. In the USA, three indices of securitised property exist: the NAREIT index, the Wilshire index and the Lehman Brothers index. As of 31 December 1994, 226 securities were included in the NAREIT index, 119 in the Wilshire index and 100 in the Lehman index. In the UK, the Financial Times–Stock Exchange Property Sector index encompasses approximately 95% of the market capitalisation of property companies. Such indices also exist in other countries, such as France (index of the Institut de l'Epargne Immobilière et Foncière (IEIF) and the Valeurs Immobilières Cotées (VIC) index, computed by ACOFI) and Switzerland (Bopp ISB A.G. index).

Several international property share indices exist: the international property share index of Datastream, the GPR indices, the Morgan Stanley Capital

International property share index and the Salomon Brothers World Equity Index–property. The GPR (Global Property Research, formerly LIFE) indices are the most comprehensive international property companies' shares indices (see Eichholtz and Koedijk, 1996a, for a comparison of these indices).

Global Property Research (GPR) computes two global real estate securities' indices: the GPR–LIFE Global Real Estate Securities index and the GPR 250 Global Real Estate Securities index. As of the end of September 1998, the GPR–LIFE Global Real Estate Securities index included 415 securities in 27 countries with a total market capitalisation of $306bn. Country and regional indices are also available. Developers are excluded from the indices, and these indices are total return indices. The number of securities and the market capitalisation of the various indices are reported in Table 4.3.

The GPR 250 Global Real Estate Securities index only includes property companies with a minimum of $50mn of freely available market value and high liquidity in terms of average last-year stock trading volume. As of December 1997, the securities included in the GPR 250 index had a combined available market value of $194bn (for details, see Eichholtz, de Graaf, Kastrop and Op 't Veld, 1998).

In order to determine if indices of indirect property can be used as benchmarks for direct investments, a critical issue is whether property companies behave like property or whether they are in fact another type of stock (see also Chapters 10 and 11). Most research has concluded that property companies' share returns are quite highly correlated with those on common stocks. Eichholtz (1997b) computes correlation coefficients between property company shares and common stocks in 19 countries, three regions and the world. The country correlation coefficients range from 0.12 (Austria) to 0.96 (Hong Kong). For Europe, the correlation is 0.82, for North America 0.71 and for the Asia/Pacific region 0.79. For the world, the correlation amounts to 0.85.

These correlation coefficients suggest that property company shares behave in a similar way to common stocks and that indices of indirect property are only very imperfect indicators of the direct property markets. However, such indices constitute useful benchmarks for investors who hold portfolios of property company shares. It is worth considering the advantages for an investor of purchasing real estate shares instead of direct property investments. First, property can be included in a portfolio even if the investor's wealth is limited, which would not be the case if direct property investments were considered. Second, with limited funds, a much better within property diversification can be achieved by purchasing shares of a property company holding a diversified portfolio. Finally, the liquidity of real estate shares is generally much greater than the liquidity of direct property investments.

In the UK and the USA, other collective equity investment vehicles are available. These vehicles are the Property Unit Trusts (PUTs) and Managed Funds in the UK and the Commingled Real Estate Funds (CREFs) in the USA. These are predominantly traded on the primary market where the price of the units is based on the regular valuations of the properties. As such, the indices of these vehicles behave like appraisal-based indices.

Table 4.3▸ Number of securities and market capitalisation for the GPR indices (as of the end of September 1998)

Country/region	Number of securities	Market capitalisation ($mn)
Argentina	1	464
Australia	26	12 723
Austria	4	585
Belgium	2	927
Canada	8	4 359
Denmark	1	297
Philippines	3	2 651
France	32	14 850
Germany	17	40 680
Hong Kong	30	27 348
Indonesia	5	63
Ireland	1	553
Italy	5	938
Japan	20	21 106
Malaysia	9	404
Norway	3	826
New Zealand	4	398
Netherlands	10	8 495
Portugal	2	231
South Africa	5	212
Singapore	12	3 534
Spain	2	1 642
Switzerland	19	5 187
Sweden	11	3 149
Thailand	1	9
UK	54	35 563
USA	128	118 347
Asia	110	68 238
Europe	163	113 922
North America	136	122 706
America	137	123 170
Asia ex Japan	90	47 131
World ex USA	287	187 195
World	415	305 542

Source: Global Property Research (GPR).

Several authors have shown that price discovery, that is information transmission, exists from the indirect to the direct property market. Such a relation has been documented for the USA, the UK, Canada and to a much lesser extent Hong Kong (see Barkham and Geltner, 1995; Eichholtz and Hartzell, 1996; Chau *et al.*, 1998). If this is indeed the case, information from the property securities markets can be used in order to predict appraisal-based indices.

This chapter has presented the types of property indices that can be used to measure the performance of property as an asset class, as benchmarks for performance evaluation purposes and for property portfolio strategy. Indices exist both for commercial property and for residential property. Owing to the scarcity of data in the commercial sector, appraisal-based indices are usually used. More data are, in most cases, available in the residential sector, making it possible to use transaction data. These data can be either unadjusted (averages of transaction prices) or quality-adjusted by the hedonic or repeat sales methods. Indices are also available for indirect property investments.

Appraisal-based indices are mostly used for commercial properties and use a sample of properties, the value of which is estimated on a regular basis. Such indices, however, are smoothed and the variance and covariance of property returns are biased towards zero (see Chapter 10). When more data are available, as is the case for residential property, transaction prices could be used. An easy way of proceeding is to compute averages of transaction prices. Limited data are required to construct such indices. This method, however, has a severe drawback in that the quality of the buildings which sell from one period to another may differ. If this is the case, the index may show a price evolution that does not reflect the true evolution of prices in the market.

A preferable approach consists in using transaction data, but controlling for the heterogeneity of the properties. The hedonic method is well suited for that purpose as it takes into account the physical and locational characteristics of the properties. Whereas hedonic indices provide an accurate reflection of price changes on property markets and are becoming widely recognised as the best way of constructing property indices, they require extensive data sets which are often difficult and onerous to assemble. Another way of controlling for the heterogeneity of property assets is to use the repeat sales method. This method uses data on properties which have sold at least twice during a given period. Several problems were identified, such as the fact that the assets which sell more than once may not be representative of the market as a whole and the fact that the quality of the assets may change between the first and the second sale.

Indices of indirect property investments also exist. They are of great use to investors in the property securities' market. Property companies behave in a similar fashion to common stocks and it is, therefore, doubtful that indices of securitised property reflect the price changes on the direct property market. Some information on appraisal-based indices can, however, be gained from lagged indirect property indices.

Notes

1. The FTSE 100 index, as its name suggests, is based on 100 shares and is recalculated every minute. It is thus used to track the evolution of prices on the market. In contrast, the All-Share index is based on approximately 650 shares and is recalculated every day only. This index is used for performance measurement analysis. For property, the IPD

monthly and JLW quarterly indices are market movement indices, while the IPD annual index is a performance measurement index.
2. Appraisal-based indices exist for commercial property in the UK, while they also exist for income-producing apartment buildings in the USA and the Netherlands, for instance.
3. A similar definition applies in the UK.
4. The 'Red Book' is the 'official' UK valuation manual for chartered surveyors. It lays down 'mandatory' procedures (which are necessary if the valuation is to be officially sanctioned and so as not to violate the valuer's personal indemnity insurance) and a set of guidelines for valuing in particular situations.
5. Retail space in the UK is divided into zones according to distance from the shop front. Standard rules apply for the size of the zones and the ratios of the rents between the different zones. Retail rents are usually quoted 'in terms of zone A' (the most expensive zone).
6. For more details on the NPI, see Pagliari *et al.* (1998).
7. The ICHP index is unusual in that it is based on properties with standardised characteristics in prime locations (whether or not such properties actually exist).
8. Prior to 1985, Morguard Investments published an index on a quarterly basis. Morguard is now a major contributor to the RCPI.
9. For details on the Australian index, see Newell and Webb (1995) and Newell and MacFarlane (1998); for the Hong Kong index, see Newell and Chau (1996); and for the New Zealand and South African indexes, see Newell and Webb (1998).
10. It is also possible to take the median price rather than the mean price.
11. The issue of temporal aggregation bias discussed for appraisal-based indices is also encountered with indices based on transaction prices.
12. A Geographic Information System (GIS) is a computerised database of area-based information. Such variables can be very useful in explaining property values. These data can be mapped to identify spatial relationships between variables. Variables in a GIS can include the noise level, shops, schools, population density and so on (see also Wyatt, 1996).

Further reading

Fisher, J. D. (1994) Real estate appraisal, in *Managing real estate portfolios*, Hudson-Wilson, S. and Wurtzebach, C. H. (Eds), Irwin, Burr Ridge (IL), 144–64: Chapter 5.
Fisher, J. D., Geltner, D. M. and Webb, R. B. (1994) Value indices of commercial real estate: a comparison of index construction methods, *The Journal of Real Estate Finance and Economics*, 9(2), 137–64.
Journal of the American Real Estate and Urban Economics Association (1991) Special issue on house prices, Haurin, D. R. and Hendershott, P. H. (Eds), 19(3).
Mark, J. H. and Goldberg, M. A. (1984) Alternative housing price indices: an evaluation, *Journal of the American Real Estate and Urban Economics Association*, 12(1), 30–49.
Morrell, G. D. (1991) Property performance analysis and performance indices: a review, *Journal of Property Research*, 8(1), 29–57.
Pagliari, J. L. Jr, Lieblich, F., Schaner, M. and Webb, J. R. (1998) Twenty years of the NCREIF property index, working draft, SSR Realty Advisors.
Shiller, R. J. (1993) *Macro markets*, Clarendon Press, Oxford: Chapter 6.
The Journal of Real Estate Finance and Economics (1997) Special issue on house price indices, Thibodeau, T. G. (Guest editor), 14(1/2).

Thibodeau, T. G. (1992) *Residential real estate prices: 1974–1983*, The Blackstone Company, Mount Pleasant.

Webb, R. B., Miles, M. and Guilkey, D. (1992) Transactions-driven commercial real estate returns: the panacea to asset allocation models?, *Journal of the American Real Estate and Urban Economics Association*, 20(2), 325–57.

Part II
Pricing models and property portfolio strategy

Chapter 5

Pricing models

Introduction

This chapter considers the application of discounted cash flow (DCF) techniques to determine whether an investment is correctly priced. It develops the concepts of the *present value* and the *internal rate of return*, which were introduced in Chapter 2. The model developed in the next sections links a number of aspects of investment – forecasting, risk and depreciation – which are considered, respectively, in Chapters 6, 3 and 7, and 8. Its application to the construction of investment portfolios is considered in Chapter 9.

The basic procedure developed below takes explicit account of the cash flows and the risk of an investment and is equally applicable at the portfolio level and the individual property level. It is in contrast with the valuation techniques used in the property market which are based on direct capital comparison of the price of recently sold and similar properties. The capitalisation rates (the initial or income yields) for these *comparable* properties are adjusted for application to the subject property. Such a procedure may be termed implicit. Nonetheless, embedded within such a valuation are assumptions about income growth, risk and depreciation.

Central to the use of a pricing model is the notion that there exist market inefficiencies, resulting in mispricing, which may be exploited. Section 5.2 considers market efficiency and mispricing, particularly in the context of the property market. Section 5.3 then derives formulae for the net present values (NPVs) of different types of cash flow. Using these formulae, it is shown how to calculate the correct yield and to compare it with the market yield and, equivalently, to calculate the expected return (the IRR) and compare it with the required return.

These two approaches are equivalent to the NPV method for determining mispricing. Section 5.4 considers the discount rate used in the DCF analyses and develops the model to show how inputs from other markets can be used to assist with the assessment of property market mispricing. The practical application of the pricing framework and the relaxation of the simplifying assumptions used are considered in Section 5.5. Finally, Section 5.6 provides a summary and conclusions.

5.2 Market efficiency and mispricing

The dominant type of fund strategy is active management in which the objective is consistently to outperform either competitors or a market benchmark. This contrasts with passive management, the dominant form of which seeks to match the market average performance in the belief that consistent outperformance is not possible (see Chapter 9 for a fuller discussion). Whatever the relative merits, active fund management dominates the investment markets. Such a strategy requires that there exists some degree of market inefficiency – that is, that market prices do not properly reflect the available information and that this mispricing may be exploited by an informed investor. The mispricing may be specific to an individual investment or systematic across a section of the market; it may be transient or may last for some time. The Efficient Markets Hypothesis (EMH) is now introduced before consideration of mispricing in the property market.

The Efficient Markets Hypothesis

The general condition for market efficiency is that prices fully reflect information quickly and accurately. If this is the case, future price changes must depend on information not yet available and so are unpredictable no matter how informed the investor is. Following Fama (1970), there are three forms of market efficiency – weak, semi-strong and strong – differing only in the information set considered:

▶ *Weak form* efficiency requires that all market information, such as prices and trading levels, is included in current price.
▶ *Semi-strong form* efficiency requires that all publicly available information, such as new product developments and financing difficulties, is included in current price.
▶ *Strong form* efficiency requires that all information is included in current price. This includes information available only to corporate insiders and specialists.

These information sets include all past information, all current information, information about events that have been announced but have not occurred and information that can be reasonably inferred. This is analogous to rational expectations and explains what, at first, may seem to be illogical price movements. Consider, for example, a share that is priced on the expectation of a company's profits falling from £10m to £5m. Suppose, instead, that profits only fall to £6m. In such circumstances, the share price would rise, despite the sharp fall in profits, because the fall was not as large as had been expected and had already been incorporated into the price.

Market efficiency does not mean that investors never make high returns, well in excess of those they require to make an investment. Rather it means that this cannot be done consistently: on average the excess return is zero.[1] Put another

way, the difference between their expected return and the actual return has a mean
of zero. If

$$\varepsilon_{i,\ t+1} = R_{i,\ t+1} - E\left(\frac{R_{i,\ t+1}}{\Phi_t}\right)$$

◄Equation 5.1

then:

$$E(\varepsilon_{i,\ t+1}) = 0$$

◄Equation 5.2

where $E(R_{i,\ t+1}/\Phi_t)$ is the expected return on asset i in period $t + 1$, conditional on
the information set Φ_t at time t, and $R_{i,\ t+1}$ is the actual return in period $t + 1$. This
can be used to test for weak form market efficiency.[2] Formal tests of property
market efficiency are limited. In a comprehensive review of the literature, Gatzlaff
and Tirtiroglu (1995) identify only five studies. Only one of these, by Brown
(1985), is for the UK. Most of the studies provide some evidence to support weak
form efficiency.

Related research provides some insights into market efficiency by examining
whether past values of prices and other variables can be used to forecast returns.
Property returns, as proxied by valuations, have been consistently shown
to exhibit *serial correlation* (Blundell and Ward, 1987; MacGregor and
Nanthakumaran, 1992; Barkham and Geltner, 1994). Serial correlation is
where returns in one period are correlated with those in previous periods
and so knowledge of past returns can help forecast future returns.

Linked to serial correlation is the notion of price discovery. Put simply, this
means that price movements in one market provide information about future price
movements in another market. Barkham and Geltner (1995) show that such a
process can be identified between the securitised (indirect) property market and the
direct property market in the USA and the UK. An extension of the analysis can
be found in the work of Chau et al. (1998) who demonstrate that there is a lagged
relationship between capital returns on a property valuation index and share
market capital returns and economic variables in Hong Kong. In essence, these
authors consider a wider range of variables that influence valuation-based returns.

The above results may seem like evidence of the failure of the property market
to conform to weak form efficiency. However, the notion of market efficiency
has been refined so that a market is defined as inefficient only if investors are
able to profit from the information, once the information search, analysis and
transactions costs, and the time to transact have been taken into account. None
of the above studies considers these issues.

Traditional valuations

Chapter 2 considered the central role of valuations in price determination in
the property market and linked this to heterogeneity, infrequent trading and
the absence of a central market. Traditional investment valuation methods in
the UK use direct capital comparison (see Chapter 4); that is, the initial yield
(capitalisation rate) from recently sold and comparable properties is subjectively
adjusted to take into account factors such as location, lease contract, tenant,

size and condition. It is then applied to the rent from the property (see Baum and Crosby, 1995, and Ball *et al.*, 1998, for a full discussion). In general, no explicit account is taken of rental growth, risk or depreciation.[3] Such valuations are assessments of the most likely selling price and assume implicitly that the market price is correct. Baum and Crosby (1995) criticise these methods as 'devoid of reality'.

These methods dominate contemporary valuation practice and may lead to mispricing. Two aspects are of importance: valuation variation and valuation accuracy. The former deals with the range of valuations for a particular property; the latter considers any systematic difference between valuations and selling price. The largest study of valuation variation is by Adair *et al.* (1996) who undertook a survey covering a range of towns and across the three main sectors in the UK and found that 202 out of 203 valuations undertaken for their study used a traditional direct capital comparison method. They grouped the valuations by sector and whether the properties were fully let or reversionary.[4] Across the groups, from 25% to 45% of valuations were greater than 10% from the average for that group. The authors acknowledge that valuers might have undertaken more research for the valuations if they had been earning a fee and that the survey was undertaken at a time of limited market transactions. Nonetheless, the study does indicate the potential for mispricing in the property market, at least for individual properties.

Valuation accuracy has been the subject of several studies. The general approach is to compare valuations with subsequent selling prices. The results tend to show a high degree of correspondence between the two but have been criticised for their lack of robustness and positive conclusions. (For a full discussion see Brown, 1992; Drivers Jonas, 1988, 1990; Lizieri and Venmore-Rowland, 1991, 1993; Matysiak and Wang, 1995; Webb, 1994.) Ball *et al.* (1998) suggest that it is possible to interpret the results as valuers acting for buyers and sellers and both being wrong. This is in line with the views of Baum and Crosby (1995) who argue that systematic mispricing occurred in the UK short leasehold market and in the over-rented City of London office market in the early 1990s and that, at least in part, it can be explained by the prevalence of traditional and inappropriate valuation techniques.

Price, valuation and worth

Given the above problems associated with valuations and their importance in price determination in the property market, it is important to distinguish among four concepts (see Baum and MacGregor, 1992; Baum *et al.*, 1996; and Ball *et al.*, 1998, for a full discussion):

▶ *Price* is the amount of money which would have to be exchanged when buying or selling a property. In most investment markets, price can be regarded as 'given' in the market, but, in the property market, where the investments are heterogeneous, it has to be estimated by a valuation. These valuations guide both seller and buyers in the negotiation of sale price.

▶ *Valuation* (sometimes termed appraisal), in the property market, is an estimate of the likely selling *price*. In contrast, in other markets, where the investments are homogeneous and price does not have to be estimated as it is determined from market trading, the term is normally used to describe an assessment of *worth*.

▶ *Individual worth* is the 'true' value of a property to an individual investor who uses all the information and analytical methods available (such as explicit cash flow analysis).

▶ *Market worth* is the price at which a property would trade in a competitive market dominated by buyers and sellers using all the information and analytical methods available.

Ball *et al.* (1998) argue that 'In an homogeneous product market, with good information and transparent transactions, market price and market worth would be equal in expectations and valuation would be a good estimate of both.' There is ample evidence to suggest that this is not true in the property market. This means that price does not always reflect the available information and, therefore, money can be made selling overpriced property and buying underpriced property. In some circumstances, the mispricing may be systematic across a section of the market. Investors who use explicit cash flow analysis and link the property market to the economy (through rental growth) and the capital markets (through the risk-free rate) can take advantage of any mispricing. The next section sets up a basic pricing model.

5.3 Building a pricing model

In Chapter 2, the expression for present value (PV) of a stream of cash flows over n periods was given as:

$$PV = \sum_{i=1}^{n} \frac{A_i}{(1 + R)^i}$$
◀**Equation 5.3**

where: PV is the present value

A_i is the amount receivable at the end of period i

R is the required return (the discount rate)

$\sum_{i=1}^{n}$ is a mathematical symbol indicating that the amounts should be

added for all values of i from 1 to n.

The net present value (NPV) is the present value, less the purchase price (P):

$$NPV = \sum_{i=1}^{n} \frac{A_i}{(1 + R)^i} - P$$
◀**Equation 5.4**

The following paragraphs develop this general model for particular types of cash flow and show how the analysis can be used to calculate the correct income yield (capitalisation rate) and compare it with the market yield or, equivalently, compare the required return with the expected return. For the present, the analysis may be considered to be either in nominal terms (with inflation included) or in real terms (with inflation removed). In the subsequent section, a distinction will be made between nominal and real analyses.

A simple cash flow model

Consider first an investment with a constant nominal income each period, D, received annually in arrears in perpetuity. This is a conventional gilt; if the cash flows were real, it would be an index-linked gilt. The present value is given by:

$$V_G = \sum_{i=1}^{\infty} \frac{D}{(1 + R_G)^i}$$

◄Equation 5.5

where: V_G is the present value, which is equivalent to the *worth* of the investment
R_G is the discount rate (the nominal required return)
$\sum_{i=1}^{\infty}$ indicates the sum to infinity.

The right-hand side is a geometric progression (that is, of the form: $a, ar, ar^2, ar^3, \ldots$) which has a sum to infinity given by:

$$\sum_{i=1}^{\infty} ar^{i-1} = \frac{a}{1 - r}$$

◄Equation 5.6

where a is the first term and r is the common ratio. In the case of Equation 5.3, the first term is $D/(1 + R_G)$ and the common ratio is $1/(1 + R_G)$. Thus, substituting in Equation 5.6 gives the sum to infinity:

$$\begin{aligned}
V_G &= \frac{D/(1 + R_G)}{1 - [1/(1 + R_G)]} \\
&= \frac{D/(1 + R_G)}{R_G/(1 + R_G)} \\
&= \frac{D}{R_G}
\end{aligned}$$

◄Equation 5.7

or

$$\frac{D}{V_G} = R_G$$

◄Equation 5.8

The true *worth* of this investment is, therefore, the constant income capitalised at the rate of required return. This worth, V_G $(= D/R_G)$, can be compared to the market *price*, P_G, to consider mispricing:

▶ If $D/R_G - P_G$ (the NPV) > 0, the investment is underpriced.
▶ If $D/R_G - P_G < 0$, the investment is overpriced.
▶ If $D/R_G - P_G = 0$, the investment is correctly priced.

A second and equivalent approach is to compare the market capitalisation rate, k_G $(= D/P_G)$, with the 'correct' capitalisation rate, D/V_G $(= R_G)$. Thus, for an investment with this simple cash flow:

▸ If $k_G > R_G$, the investment is underpriced.
▸ If $k_G < R_G$, the investment is overpriced.
▸ If $k_G = R_G$, the investment is correctly priced.

A third approach is to compare the *required return* (R_G) with the *expected return* (k_G). For a constant income, this third approach is identical to the capitalisation rate comparison.

Of the three equivalent investment decision rules provided above, the last two are easier to use.

Suppose an undated government bond is on sale with a market capitalisation rate of 7% and an investor's required return is 8%. Then, the expected return is:

$$k_G = 7\%$$

and the required return is:

$$R_G = 8\%$$

Thus the expected return is less than the required return $(k_G < R_G)$ and the gilt is overpriced.

The analysis is now extended to consider a cash flow that is increasing from period to period at a constant rate. This income profile is similar to that of a share.

Income growth and Gordon's model

Using the same notation as before, introduce a constant *expected* nominal growth rate, G_S. Thus, the cash flow is:

$$D; \ D(1 + G_S); \ D(1 + G_S)^2; \ \ldots; \ D(1 + G_S)^i, \ \ldots,$$

and the value is:

$$V_S = \sum_{i=1}^{\infty} \frac{D(1 + G_S)^{i-1}}{(1 + R_S)^i} \qquad \qquad \blacktriangleleft \textbf{Equation 5.9}$$

where: V_S is the present value, which is equivalent to the worth of the investment
G_S is a constant growth rate for income
R_S is the discount rate.

The right-hand side is again a geometric progression, this time the first term is $D/(1 + R_S)$ and the common ratio is $(1 + G_S)/(1 + R_S)$. Substituting in Equation 5.6 gives the sum to infinity:

$$V_S = \frac{D/(1 + R_S)}{1 - [(1 + G_S)/(1 + R_S)]}$$

$$= \frac{D/(1 + R_S)}{(R_S - G_S)/(1 + R_S)}$$

$$= \frac{D}{R_S - G_S}$$

◄Equation 5.10

or

$$\frac{D}{V_S} = R_S - G_S$$

◄Equation 5.11

This is known as Gordon's growth model and is widely used in the share market. The true worth of a share is the initial income, capitalised at the rate of the required return less the expected income growth rate. V_S $(= D/(R_S - G_S))$ can be compared to the market price, P_S, to determine mispricing:

▸ If $D/(R_S - G_S) - P_S$ (the NPV) > 0, the investment is underpriced.
▸ If $D/(R_S - G_S) - P_S < 0$, the investment is overpriced.
▸ If $D/(R_S - G_S) - P_S = 0$, the investment is correctly priced.

As for the constant cash flow above, an equivalent approach is to consider the market capitalisation rate, k_S $(= D/P_S)$, and to compare this with the 'correct' capitalisation rate, D/V_S $(= R_S - G_S)$:

▸ If $k_S > R_S - G_S$, the asset is underpriced.
▸ If $k_S < R_S - G_S$, the asset is overpriced.
▸ If $k_S = R_S - G_S$, the asset is correctly priced.

By manipulating the above rule, it is also possible to compare the required return (R_S) with the expected return $(k_S + G_S)$:

▸ If $R_S < k_S + G_S$, the asset is underpriced.
▸ If $R_S > k_S + G_S$, the asset is overpriced.
▸ If $R_S = k_S + G_S$, the asset is correctly priced.

Note that if G_S is other than expected, the actual return will differ from that expected.

Consider a share with a market capitalisation rate of 6%. The investor has a required return of 10% and expects the share to have income growth of 5%. Then, the expected return is:

$$k_S + G_S = 6\% + 5\% = 11\%$$

and the required return is:

$$R_S = 10\%$$

Thus the expected return is greater than the required return $(k_G + G_S > R_G)$ and the share is underpriced.

A knowledge of the required return and a view on expected income growth enables an investor to determine the correct capitalisation rate. To produce a basic property-pricing model requires the introduction of one further variable, depreciation.

Depreciation

Depreciation is covered in detail in Chapter 8 and is introduced here only briefly. It may be thought of as the loss of rental income (and hence of market value) of an ageing building when compared to an equivalent new building. This results from wear and tear, the effects of the elements and changing functional and locational requirements of occupiers. In some cases, expenditure to refurbish the property may result in an increase in rent. However, this expenditure can be considered to be the equivalent of a future stream of costs which reduces future rent. Either the actual lower income or the notional lower income provides a measure of depreciation when compared to the rent from an equivalent new property.

The depreciation rate, d, is taken as a constant, just as the discount rate and income growth were in the analysis above. There are two ways to consider depreciation: as a loss of income or as a discount rate.

Depreciation as a loss of income

Assume a rental income, D, for new buildings, growing at a constant rate, G_P. Then a continuously new building will have a rental value of $D(1 + G_P)^n$ after n periods.[5] The effect of depreciation is a geometric decrease in rent each period *when compared to the new building*. Thus the income stream is:

$$D; \quad D(1 + G_P)(1 - d); \quad D^2(1 + G_P)^2(1 - d)^2; \quad \ldots; \quad D(1 + G_P)^i(1 - d)^i; \quad \ldots$$

This method has the advantage that, say, 5% depreciation results, after one year, in a rent for the ageing building of 95% of the rent of a new building. Thus, 5% depreciation results in a 5% loss of rent.

Depreciation as a discount rate

An alternative method is to consider depreciation as a further discount factor. The income stream is:

$$D; \quad D(1 + G_P)/(1 + d); \quad D(1 + G_P)^2/(1 + d)^2; \quad \ldots; \quad D(1 + G_P)^i/(1 + d)^i; \quad \ldots$$

In this case, 5% depreciation does not result in a 5% loss of rent but in a loss of $(1 - [1/(1 + d)]) = 4.77\%$. However, this method has the advantage that, say, 5% rental growth and 5% depreciation would cancel out to give a constant rent.

The difference between the two is insignificant for likely values of d which have been estimated to be in the range of 0–5% depending on the type of property.[6] For the calculations below, the first method will be used, but either produces the same approximation.

A property-pricing model

Assume, as above, an initial income, D, paid annually in arrears, with a constant rate of nominal income growth, G_P, and a constant rate of depreciation, d, and a constant nominal discount rate, R_P. For the moment, it is also necessary to assume annual rent reviews. Thus:

$$V_P = \sum_{i=1}^{\infty} \frac{D(1 + G_P)^{i-1}(1 - d)^{i-1}}{(1 + R_P)^i} \qquad \blacktriangleleft \text{Equation 5.12}$$

where V_P is the present value, which is equivalent to the worth of the investment.

The right-hand side is a geometric progression with a first term of $D/(1 + R_P)$ and a common ratio of $(1 + G_P)(1 - d)/(1 + R_P)$. Substituting in Equation 5.6 gives the sum to infinity:

$$V_P = \frac{D/(1 + R_P)}{1 - [(1 + G_P)(1 - d)/(1 + R_P)]}$$

Thus:

$$V_P = \frac{D/(1 + R_P)}{[(1 + R_P)/(1 + R_P)] - [(1 + G_P)(1 - d)/(1 + R_P)]}$$

$$= \frac{D}{(1 + R_P) - (1 + G_P - d - G_P d)}$$

$$= \frac{D}{R_P - G_P + d + G_P d}$$

As G_P is likely to take values well under 10% and d is under 5%, the final term, $G_P d$, is very small and may be ignored. Thus:

$$V_P = \frac{D}{(R_P - G_P + d)} \qquad \blacktriangleleft \text{Equation 5.13}$$

or

$$\frac{D}{V_P} = R_P - G_P + d \qquad \blacktriangleleft \text{Equation 5.14}$$

The true worth of a property investment is the initial income, capitalised at the rate of the required return, less the expected income growth rate of a new property, plus the expected depreciation. $V_P \, (= D/(R_P - G_P + d))$ can be compared to the market price, P_P, to determine mispricing:

▸ If $D/(R_P - G_P + d) - P_P$ (the NPV) > 0, the investment is underpriced.
▸ If $D/(R_P - G_P + d) - P_P < 0$, the investment is overpriced.
▸ If $D/(R_P - G_P + d) - P_P = 0$, the investment is correctly priced.

Alternatively, the market capitalisation rate, $k_P \, (= D/P_P)$, can be compared to the 'correct' capitalisation rate, $D/V_P \, (= (R_P - G_P + d))$. This again produces a simple rule to determine mispricing:

▸ If $k_P > (R_P - G_P + d)$, the asset is underpriced.
▸ If $k_P < (R_P - G_P + d)$, the asset is overpriced.
▸ If $k_P = (R_P - G_P + d)$, the asset is correctly priced.

By manipulating the above rule, the required return (R_P) can be compared with the expected return ($k_P + G_P - d$):

- If $R_P < k_P + G_P - d$, the asset is underpriced.
- If $R_P > k_P + G_P - d$, the asset is overpriced.
- If $R_P = k_P + G_P - d$, the asset is correctly priced.

Note that the actual return will differ from the expected return if G_P or d differs from its expected value.

Consider a property with a market capitalisation rate of 8%. The investor has a required return of 9%. Income growth is expected to be 3% and depreciation is expected to be 2%. Then, the expected return is:

$$k_P + G_P - d = 8\% + 3\% - 2\% = 9\%$$

and the required return is:

$$R_S = 9\%$$

Thus, the expected return is equal to the required return ($k_P + G_P - d = R_P$) and the property is correctly priced.

Note that, unlike the simple model and the income growth model, the formula here is an approximation rather than exact.[7] However, for likely values of the inputs, the approximation is highly accurate.

The next section considers the discount rates appropriate for undertaking the analysis in nominal and real terms. It also shows how information from other markets can be used to consider property mispricing.

5.4 The discount rate

Although it is conventional in property investment to use a nominal framework for analysis, the DCF analysis outlined above can be undertaken either using nominal cash flows and a nominal discount rate, or real cash flows and a real discount rate. Forecasting models typically produce real values for rents as these are linked to some measure of real activity in the economy (see Chapter 6). However, the forecasts are of market rents and not of rents receivable which, in the UK, are fixed for five years in nominal terms. Accordingly, for property, a real analysis still requires forecasts of inflation to determine the real value of the fixed nominal rent (see below).

Chapter 2 introduced the discount rate as the required return. An investor must expect to receive this in order to make an investment. The required return comprises two parts: a risk-free rate and a risk premium. The first is common to all investments while the second varies from investment to investment depending on its risk. If a nominal analysis is used, the nominal required return is:

$$R_N = RF_N + RP$$

◄Equation 5.15

where: R_N is the required nominal return
   RF_N is the nominal risk-free rate
   RP is the risk premium.

In this case, the risk-free rate should have no nominal risk, that is, there should be a guaranteed nominal return. As shown in Chapter 2, this is provided by the gross redemption yield on conventional gilts. These should be of the same length as the investment under consideration. If the investment is analysed in perpetuity, the appropriate risk-free rate is the gross redemption yield (the IRR) on undated gilts.

The analysis can also be undertaken in real terms. This means that inflation is removed from the cash flows, a real risk-free rate and a real required return are used and the return calculated is a real return. The risk-free rate has to be one with no real risk, in other words, there is a guaranteed real return. This is the gross redemption yield on index-linked gilts as all the cash flows are known in advance in real terms, so the gross redemption yield (the IRR) provides a real risk-free return.[8]

Chapter 2 also showed that the conventional gilt yield contains within it the market's expectation of inflation. Thus, a conventional gilt is inflation prone as, if there is unexpected inflation, the real return achieved will be different from that expected. As real returns are important to most investors, this means a conventional gilt has inflation risk which must be rewarded. The relationship between the risk-free nominal and real rates is:

$$RF_N = RF_R + iRP + i_e \qquad\qquad \blacktriangleleft\text{Equation 5.16}$$

where: RF_N is the nominal risk-free rate
   RF_R is the real risk-free rate
   iRP is an inflation risk premium
   i_e is expected inflation.

Combining Equations 5.15 and 5.16 gives:

$$R_N = RF_R + RP + iRP + i_e \qquad\qquad \blacktriangleleft\text{Equation 5.17}$$

If the inflation risk premium is included in the general risk premium, this is the Fisher equation, introduced in Chapter 2:

$$R_N = RF_R + RP + i_e \qquad\qquad \blacktriangleleft\text{Equation 5.18}$$

The last of these, expected inflation, is common to all investments. The first is conventionally taken as the same for all assets on the assumption of high liquidity. If, as in the property market, there is significantly lower liquidity, it is appropriate to add a differential liquidity premium. For ease of exposition, this can be subsumed within the risk premium which becomes a risk and liquidity premium. This final component, the risk premium, depends on the risk of the investment.

Comparing the property market with other investment markets

The analysis of price and worth (in nominal terms) can be extended to link it with other investment markets. Equation 5.14, for the correct initial yield, can be expanded as:

$$\frac{D}{V_P} = RF_R + RP_P + i_e - G_P + d \qquad \qquad \blacktriangleleft\textbf{Equation 5.19}$$

If the market is correctly priced:

$$k_P = \frac{D}{V_P} = RF_R + RP_P + i_e - G_P + d \qquad \qquad \blacktriangleleft\textbf{Equation 5.20}$$

This may be rewritten as:

$$RF_R + RP_P + i_e = k_P + G_P - d \qquad \qquad \blacktriangleleft\textbf{Equation 5.21}$$

This compares the required nominal return ($RF_R + RP_P + i_e$) with the expected nominal return ($k_P + G_P - d$). For a correctly priced market, the two are equal. The required return comprises a real risk-free return, a risk premium and expected inflation; the expected return comprises an initial return, plus expected nominal income growth less depreciation.

These formulae may be used to compare pricing in the different asset markets and to analyse mispricing, as in Example 5.4.

Assume that the initial yields (capitalisation rates) on the main asset classes are as follows:

- index-linked gilts (ILGs): 3.0%
- conventional gilts: 7.5%
- shares: 3.5%, and
- property: 7.0%.

Use this information to compare the nominal required return with the expected return in each market.

This can be done in a number of stages (see Table 5.1). It involves drawing inferences about the components of required and expected return:

1. The capitalisation rate for ILGs provides the real risk-free rate which is common to all assets. Thus, $RF_R = 3.0$ for all assets.
2. The difference between the yields on conventional gilts and ILGs comprises the market's expectation of inflation (i_e) for the life of the gilt plus an inflation risk premium. Taking an inflation risk premium of 0.5%, gives expected inflation of 4.0%. This is common to all assets.
3. Depreciation (d) is zero except for property (see Chapter 8). The value varies according to property type and is highest for industrials and lowest for retail. For property in the aggregate, a value of around 1% is about right.
4. By definition, the risk premium (RP) for index-linked gilts is zero.
5. The delivered risk premium for shares has been about 2.5%.[9] Property has lower volatility in returns (see Chapter 10) and benefits from upward-only rent reviews. However, it suffers from illiquidity. Taken together, these suggest a premium of around 3.5%.
6. Index-linked gilts have income that rises in line with actual inflation, so expected nominal income growth (G) is the same as expected inflation: 4.0%. Conventional gilts have a constant nominal income, so zero growth.

The growth figures for property and shares may then be inserted to ensure that the required return $(RF_R + RP_P + i_e)$ equates to the expected return $(k + G - d)$ – that is, starting from the assumption of correct pricing.

Inspection of the growth rates shows share income rising at 2.0% in real terms (6.0% nominal) and property income at 0.5% real (4.5% nominal). Both of these are broadly in line with historical averages in the UK. Note that if the conventional 2% risk premium is used for property, a nominal income growth rate of 3.0% (−1.0% real) is implied. This is difficult to justify. However, a property risk premium of 3% would imply real growth of zero – that is, nominal growth in line with inflation. This would be a normal expectation in the USA where the planning system allows supply to meet demand.

If any possible mispricing had been identified, it would be appropriate first to consider other values for the risk premia which would allow values for growth more in line with long-term averages. If this were not possible, further detailed consideration would be required, as in Section 5.5 below.

Table 5.1▶ Using other asset markets to consider property mispricing

	Required return				Expected return			
	RF_R	+RP	+i_e	Total	k	+G	−d	Total
Index-linked gilts	3.0	0.0	4.0	**7.0**	3.0	4.0	0.0	**7.0**
Conventional gilts	3.0	0.5	4.0	**7.5**	7.5	0.0	0.0	**7.5**
Shares	3.0	2.5	4.0	**9.5**	3.5	6.0	0.0	**9.5**
Property	3.0	3.5	4.0	**10.5**	7.0	4.5	1.0	**10.5**

The figures presented in Example 5.4 are for illustration only and refer to the UK. Although it is beyond the scope of this chapter to provide figures for several countries it must be stressed that these figures vary from country to country.[10]

The basic model presented above can be used at all levels of analysis: for all property; property sectors; sectors within a region; towns or individual properties. However, it is a simple and crude analysis which needs to be refined for use in actual investment decision-making. The next section builds up a practical pricing model which allows buy and sell decisions to be made.

5.5 Applying the procedure in practice

The procedure developed above contains two simplifying assumptions – the pattern of cash flows and constant growth rates – which must be relaxed if it is to be used in practice.[11]

Cash flows

The income from gilts (government bonds) is received twice yearly in arrears in equal parts; for shares, income is received twice yearly in unequal parts; and for property in the UK income is received quarterly in advance.[12] As shown in Example 5.5 and Table 5.2, while the difference is relatively small for bonds and shares, it is significant for property even for small values of the discount rate.

Consider three patterns of cash flow for £100 received during a year: annually in arrears (the standard financial mathematics assumption); twice annually in arrears (shares and bonds); and quarterly in advance (UK property).

Table 5.2► The present value of £100

Discount rate	Annually in arrears	Twice yearly in arrears	Quarterly in advance
6%	£94.34	£95.73	£97.85
8%	£92.59	£94.41	£97.18
10%	£90.91	£93.13	£96.52
12%	£89.29	£91.89	£95.89

Note: Assumes equal payments.

A further complication is that, in the UK, property income is typically fixed for five years between rent reviews and then adjusted (upwards only) to the open market rent.[13] This means, as shown in Example 5.6 and Table 5.3, that the income stream is less valuable than one which is reviewed annually or twice yearly.[14] Over a five-year period, even at low levels of income growth, this loss far outweighs the advantage of quarterly-in-advance payments. Thus, when applying the basic model to the assessment of mispricing in practice, it is essential to incorporate the actual pattern of cash flows into the analysis.

Consider £100 received annually for five years with differing payment patterns. The property income is not reviewed during the five-year period.

Table 5.3► The present value of an initial income of £100 a year with varying growth rates for five years

Growth rate	Annually in arrears	Twice yearly in arrears	Quarterly in advance (without review)
0%	£379.08	£388.33	£402.49
4%	£415.66	£421.57	£402.49
5%	£425.31	£430.31	£402.49
6%	£435.16	£439.23	£402.49
7%	£445.23	£448.33	£402.49

Note: Assuming a discount rate of 10% and equal payments.

Short-term forecasts

So far it has been assumed that the rate of growth in the nominal cash flow is constant. Relaxing this assumption allows more realistic analysis.

One of the central notions in economics is that economic systems tend towards equilibrium.[15] For example, when an economic variable, such as real output, is above or below its long-term trend, it will tend to revert to the trend.[16] Thus, when growth rates are above average, they will tend to fall; and if they are below average, they will tend to rise. This is also known as *mean reversion* and is empirically observable in economic variables.

Income from real assets is linked to economic performance so, it too, has a long-term trend. The long-term trend in real rents is linked to demand and supply.

Thus, real rents can be considered as having short-term fluctuations around this long-term trend. Whether it is above or below the trend, depends on the state of the economy and the stage of the economic and property cycles.[17] Short-term forecasting models may be built to link rental growth to the appropriate economic variables (see Chapter 6). These short-term forecasts of rental growth may be used in the cash flow analysis of mispricing. Beyond these forecasts, the long-term averages may be used. The question arising is 'How long is short term?'

It is not possible to produce a sensible forecast of real rental growth for the year 2020 but it is possible to estimate the average level of growth from 2010 to 2030 based on the long-term average. On the other hand, it is possible to have a view on real rental growth next year based on the forecasts for the economy for next year. Such economic forecasts are available and have some value for up to five years, so it is possible to produce year by year property rental growth forecasts for five years. Thus, rather than assume constant real rental growth, it is possible to have explicit short-term forecasts of rent for up to five years. Beyond that period, year by year forecasts have little value and so a long-term average figure has to be used.

The model in practice

The adjustments to the cash flow patterns and the introduction of short-term forecasts require that the model is designed on a spreadsheet, although the design is straightforward. The adjustments permit assessments to be made of mispricing using the inputs of the market yield (capitalisation rate) and the investor's assessment of income growth, risk and depreciation.

Consider the analysis of correct price for the inputs shown in Table 5.4.

Table 5.4► Inputs and outputs for a pricing analysis of the UK office sector

Inputs

Year	Real rental growth (g_P) (1)	Expected inflation (i_e) (2)	Nominal rental growth ($G_P = (1 + g_P)(1 + i_e) - 1$) (3)	Depreciation (d) (4)
1	−3.0%	4.0%	0.9%	0.5%
2	2.0%	4.0%	6.1%	0.5%
3	4.0%	4.5%	8.7%	0.5%
4	4.0%	6.0%	10.2%	0.5%
5	5.0%	4.5%	9.7%	0.5%
5+	0.0%	3.5%	3.5%	0.5%

Current yield (k_P)	7.0%		
Real risk-free rate (RF_R)	3.5%	} Real required return ($RF_R + RP_P$)	6.5%
Risk premium (RP_P)	3.0%		

Outputs (based on price of £100)

Nominal IRR	12.76%		
Real IRR	7.84%	Present value	£105.5
Required real return	6.50%	Net present value	£5.5
Excess real return	1.34%	Correct yield	6.64%

Note: The IRRs are calculated from the nominal and real cash flows constructed from the panel above. The correct yield is the current income (£7) divided by the present value (£105.5).

Example 5.7 and Table 5.4 illustrate model inputs and outputs for an analysis of the office sector, assuming that rents have just been reviewed to open market levels. The short-term real rental growth forecasts (column 1) are derived from a forecasting model linking forecasts of key economic variables to rents. The inflation forecasts (column 2) are from economic forecasts. Long-term averages are used beyond five years. Together these produce forecasts of nominal rental growth (column 3). Depreciation estimates (column 4) are taken from published analyses. The current yield is market information; the real risk-free rate is taken from the index-linked gilt market; and the risk premium is an assessment based on research and the investor's risk intolerance. Thus, the model links the property market directly to the economy and to the capital markets.

There are three main outputs. First, the estimated nominal and real cash flows are used to calculate the nominal and real internal rates of return (IRR). When the required real return is subtracted from the latter, the excess real return is produced. Second, the net present value (NPV) is based on an initial purchase price of £100. Third, the correct yield (capitalisation rate) is calculated from the current income (£7: 7% of the price of £100) and the present value (£105.5). Note that these three analyses always produce the same result. If the excess return is positive/negative/zero, the NPV will be positive/negative/zero and the correct yield will be below/above/equal to the market yield.

Fundamentally, this analysis assumes a five-year holding period during which yields will adjust to the correct level and the market will return to equilibrium. The terminal capital value in the analysis (at five years) is the equilibrium value and the IRR is an average figure for the five-year period. It is possible to calculate the correct yield each year and to impose a pattern of adjustment over time to the correct yield. This permits the calculation of year-by-year returns. Such an analysis may be based on past adjustment patterns but is likely to be speculative.

In practice, there will be another shock and further adjustments to equilibrium. This does not reduce the value of the approach but it does require some caution in practical implementation. This basic framework is used in the share market where information is typically much more freely available and research rather more developed.

Even for informed investors using this analysis, the inputs will vary depending on their risk assessment, and on their estimates of expected inflation, real rental growth and depreciation. Views on the risk of an investment affect the discount rate used in the model.

Real rent and inflation forecasts are used in the cash flow analysis. A number of complex econometric models exist of the UK economy and a range of economic forecasts is produced. Any one of these could be used to provide inflation forecasts and growth forecasts as inputs into a rent forecasting model (see Chapter 6). Different rent models will produce different outputs from the same inputs. Finally, depreciation has been inadequately researched and is still not widely understood or incorporated explicitly into pricing analyses (see Chapter 8).

Thus, a range of views is likely for the property market as a whole and for individual sectors, and a wider range might be expected at town or individual property level. While the range of inputs provides different views on worth in the property market, the scope for mispricing is exacerbated by the relatively poor data available, the limited amount of formal forecasting undertaken below the portfolio level and the continuing prevalence of implicit traditional valuation procedures. This means that mispricing, whether for an individual property or systematically across a market section, is possible. Any mispricing can be identified and exploited by an investor using explicit cash flow analysis and with access to good quality research. The use of the outputs from the model in portfolio strategy is considered in Chapter 9.

5.6 Summary and conclusions

This chapter has shown how a pricing model can be used to identify and exploit mispricing in the property market. The notion of market efficiency was considered in the context of property. Three forms of efficiency were introduced: strong, semi-strong and weak. There is some limited evidence to suggest that the property market may conform to weak form efficiency. On the other hand, the existence of serial correlation in returns and of price discovery between the securitised and the direct property markets suggest that mispricing may exist. The issue, then, is whether it can be exploited profitably or whether cost and time would prevent this.

The dominance of traditional implicit valuation techniques was argued to create substantial potential for mispricing, either specific to an individual property or systematic across a section of the market. Research shows that valuers using such techniques produce a wide range of valuations for the same properties. Although there is some evidence to suggest that, on average, valuers are able to estimate the subsequent selling price, such results have been challenged. There appears clearly to be scope for mispricing of individual properties, and some authors have argued that such mispricing has been systematic in the UK, lasting significant periods for particular types of property such as short leaseholds and over-rented offices. In such a market, it is important to distinguish price, valuation, individual worth and market worth.

Discounted cash flow analysis was used to produce three equivalent decision rules for investments with different cash flows. The rules are based on the net present value (NPV), comparison of the market capitalisation rate with the correct rate, and comparison of the required return with the expected return. These rules were derived for investments with a constant income (equivalent to a bond), with constant income growth (a simplified version of a share) and with constant income growth and depreciation (a simplified version of a property). The rules provide basic tests of mispricing.

The analysis was then extended to consider both nominal and real frameworks. Nominal and real-risk free rates, derived from the gilt markets, were compared and introduced into the analysis as part of the required return. This required return, which can be decomposed into a risk-free rate and a risk premium, is the discount rate used in the cash flow analysis. The property market capitalisation rate can be compared with those in other capital markets to provide inputs into an analysis of mispricing.

The model was then developed in two ways. First, the actual pattern of UK property cash flows -- quarterly in advance with rents fixed for five years between reviews – was introduced rather than the simplifying annual in arrears assumption. Second, in line with the notion of economic systems tending to return to equilibrium after a shock, the assumption of constant income growth was relaxed to allow variable rates for the first five years. These are based on short-term economic forecasts and rent-forecasting models linking the property market to the economy.

The model requires forecasts of rents, depreciation rates, analysis of risk and a risk-free rate. It links the property market directly to the economy and to the capital markets. Such analysis is beyond the resources of most investors in property, which, in the context of a market dominated by traditional implicit valuation methods, suggests that there are opportunities to exploit mispricing. It may be difficult to assemble the inputs, but it can be done. Subsequent chapters consider the main inputs – forecasts, depreciation and risk – and the use of the outputs in practical portfolio management.

Notes

1. More strictly, it is not possible to outperform the market consistently, except by taking above-average risk. In such circumstances, the risk-adjusted excess return would not be consistently above zero.
2. Tests for strong and semi-strong efficiency are more difficult (see, for example, Adams, 1989).
3. Variants of the basic methods derive the market's expected rental growth and apply that to the subject property. These suffer from two problems: first, the value for the growth rate depends on the required return used and this is rarely derived from the capital markets; and, second, no attempt is made to assess whether this is realistic based on the economic factors affecting growth. In short, the analysis views the property market in isolation.
4. A fully let property has a passing rent at the open market rent; a reversionary property has a passing rent below the open market rent.
5. There is, of course, no such thing as a continuously new building but it is a useful device for illustration.
6. Note that $1/(1 + d)$ can be expanded as an infinite series as: $1 - d + d^2 - d^3 + d^4 - \ldots$, so for small values of d, the two approaches are equivalent.
7. Note that if *continuous* cash flows are assumed, the infinite summation for V_p becomes the integral

$$\int_0^\infty De^{-(R-G_p+d)t}dt = \frac{D}{(R - G_p + d)}$$

In this case, the formula is exact.
8. This, of course, cannot be established in countries with no index-linked government bonds.
9. In recent years, and particularly in the USA, the delivered risk premium has been much higher. Nonetheless, the figure of 2.5% is used here to illustrate the general method.

10. For example, the prime office investment yield in 1997 was 5% in London, 6% in Paris, 6.5% in Amsterdam and 7% in Brussels. For New York and Sydney the figures were 8% and 6.5%, respectively (Richard Ellis, 1997). The annual depreciation rate on commercial property in the USA is 2.5% (Brueggeman and Fisher, 1997, 339). In France the ratio of rent to value (the capitalisation rate) is in the 6.5–10% range for offices and shops in city centres and in the 8–11% range for shopping centres. The annual depreciation rate is 1–3% for offices and shops in city centres and 2–4% for shopping centres (Dumortier, 1997: 14).
11. Other elaborations are time-varying discount and depreciation rates, but these are not considered here.
12. In most countries rents are paid monthly in advance (for instance, in the USA and in Germany).
13. This, in effect, creates an 'option' to hold rents constant. The option would be exercised if open market rents were to fall below the contract rent. Other lease terms also create options. In the USA, turnover (overage) rent clauses are common in retail leases. If sales rise above a specified level, rent is increased by a proportion of the amount above that level (usually from 1% to 10%). These options have a value and should be taken into account when determining worth. Such analyses have been applied to lease terms only recently: see Grenadier (1995) and Ward et al. (1998). As the subject is not well developed in the property market, and as lease terms vary so much from country to country, it is not considered here.
14. In most countries rents are reviewed annually. See also Chapters 2 and 12.
15. For a review of the use of equilibrium models in property markets, see Hendershott (1997).
16. A shock can alter the long-term trend but this merely means that there is a new equilibrium.
17. For a full discussion of the property cycles, see, for example, Ball et al. (1998).

Further reading

Adair, A., Downie, M. L., McGreal, S. and Vos, G. (Eds) (1996) *European valuation practice: theory and techniques*, E & FN Spon, London.

Ball, M., Lizieri, C. M. and MacGregor, B. D. (1998) *The economics of commercial property markets*, Routledge, London.

Baum, A. E. and Crosby, N. (1995) *Property investment appraisal*, London, Routledge (second edition).

Baum, A. E., Crosby, N. and MacGregor, B. D. (1996) Price formation, mispricing and investment analysis in the property market, *Journal of Property Valuation and Investment*, 14(1), 36–49.

Baum, A. E. and MacGregor, B. D. (1992) The initial yield revealed: explicit valuations and the future of property investment, *Journal of Property Valuation and Investment*, 10(4), 709–27.

Brueggeman, W. B. and Fisher, J. D. (1997) *Real estate finance and investments*, Irwin, Chicago (tenth edition): Chapter 10.

Corgel, J. B., Smith, H. C. and Ling, D. C. (1998) *Real estate perspectives: an introduction to real estate*, Irwin McGraw-Hill, Boston (third edition): Chapter 3.

Gelbtuch, H. C., Mackmin, D. and Milgrim, M. R. (Eds) (1997) *Real estate valuation in global markets*, Appraisal Institute, Chicago (IL).

Hendershott, P. H. (1997) Uses of equilibrium models in real estate research, *Journal of Property Research*, 14(1), 1–13.

Ross, S. A., Westerfield, R. W. and Jaffe, J. (1993) *Corporate finance*, Irwin, Chicago (fourth edition): Chapter 7.

White, P. (1995) A note on 'The initial yield revealed: explicit valuations and the future of property investment', *Journal of Property Valuation and Investment*, 13(3), 53–8.

Wurtzebach, C. H. and Miles, M. E. (1991) *Modern real estate*, John Wiley & Sons, Inc., New York (fourth edition): Chapter 21.

Chapter 6

Forecasting property markets

6.1 Introduction

This chapter considers methods of forecasting property markets. Such forecasts are an essential input into the pricing model of Chapter 5 and, consequently, into the development of a portfolio strategy (Chapter 9). In Chapters 2 and 5 it was shown that the correct price for an investment depends, in part, on its expected income growth. Other things being equal, the higher the expected income growth, the lower the capitalisation rate and the higher the value. Thus, when an investment decision is made to buy or sell, whether it is based on a cash flow assessment of worth or a traditional valuation involving the capitalisation of current income at the all risks yield, a forecast of rental growth is involved. In the former case it is an explicit part of the cash flow analysis; in the latter, the forecast is implicit and is embedded within the capitalisation rate.

The formal forecasting of the UK property market using econometric models grew rapidly from the mid-1980s, although cruder (and often misleading) correlation and regression analyses were undertaken from the late 1970s. There are a number of reasons for this development:

1. The 1980s saw a dramatic expansion of property research to meet the needs of major investors who sought to adopt a portfolio perspective for the management of their property assets and to link the property market to the economy and to the capital markets (see Chapters 5 and 9). One fundamental requirement was rental growth forecasts for use in explicit cash flow analyses to assess pricing and to produce return forecasts for portfolio construction.
2. The expansion of research brought new skills from economics and finance to the property market.
3. By the mid-1980s, there was available a long enough time series of rents to make the building of simple models a realistic possibility.
4. The development of cheap, user friendly and powerful econometrics software for PCs enabled easier analysis.

As a consequence there has been a dramatic increase in the availability of forecasts and in the quality of the models from which forecasts are derived. A number of companies in the UK now offer a combination of forecasts and portfolio strategy advice, and most larger investors and surveying firms employ in-house researchers to produce or interpret forecasts.

As many of the developments in forecasting have been in the private sector and the results are commercially sensitive, there has been, until recently, limited published material on the subject. This chapter draws on the published literature and covers the basic steps involved in building a model but does not propose definitive models. It deals with models which link property market variables to economic and capital market variables. A basic knowledge of statistics, in particular regression analysis, is assumed.

Simple trend analysis which might, for example, estimate average rental growth over a period and extrapolate this growth as a forecast, is not covered. Such an approach has no value for shorter term forecasting although it is useful for estimating longer term average growth rates for use in yield analysis (see Chapter 5). Nor does this chapter deal with time series models which link the future value of rental change only to its past values. Some such models are of value in forecasting one or two periods in quarterly data but are of little value in producing the medium term forecasts considered here. In any case, the mathematics of all but the most basic of these is beyond the scope of this book.

Section 6.2 sets out a basic framework for model building in the property market, including some of the statistical principles involved. Section 6.3 starts by considering the economic principles underlying a simple model which links rent to the demand for, and supply of, property. The retail sector is used to illustrate the modelling issues. It also considers the simpler rental adjustment models common in the USA, particularly for the office market. Next the simple model is expanded to show how demand and supply equations can be estimated separately to produce a three-equation model. This is illustrated using the office market. Section 6.4 discusses the extension of the simple forecasting model framework to the regional and town level. It shows that, while regional modelling can follow the same basic principles as national level modelling, town level modelling is fraught with conceptual and practical difficulties. Section 6.5 then considers the difficulties in forecasting capitalisation rates and total returns. Finally, section 6.6 provides a summary and conclusions.

6.2 Model building

Introduction

The type of forecasting model under consideration in this chapter is one which links the variable to be forecast, known as the *dependent* variable, to one or more other variables, known as *independent* or *explanatory* variables. Changes in the independent variables are used to explain changes in the dependent variable. Building the model involves estimation of the relationships between the dependent variable and appropriate independent variables from historical data. This section sets out the basic procedures involved. It considers conceptualisation of the model, preliminary data analysis and preparation, model specification, model estimation, model testing and forecasting. The material presented is relatively basic and, while a willingness to deal with equations is required, no statistical knowledge is necessary to be able to follow the procedure. It should be stressed that these are not a series of discrete stages to be considered one after the other but, instead, a general ordering of the approach: some amount of iteration among stages is usually necessary and is certainly advisable. Application to actual data requires access to decent econometrics software, and a number of powerful and user friendly packages are available. As in all such analyses, it is essential to understand the basics to be able to build sensible models. In Section 6.3, the issues are illustrated in practice using the example of the UK property market.

Conceptualisation of the model

To ensure that the model produced is both logical and plausible, the procedure should start with consideration of the relevant theory, in this case economic theories of the operation of the property market, to help to identify appropriate explanatory variables. The best model is likely to be simple and to contain a small number of explanatory variables. While adding new variables always improves the statistical fit of a model (measured by R^2, see below), a large number of explanatory variables is likely to stretch the plausibility of the underlying theory. In property market model building, plausibility usually goes hand in hand with parsimony and elegance.

While theory may suggest appropriate explanatory variables, these are often no more than economic concepts. To build an empirical model requires that it is possible to measure these concepts directly or by proxy. Such considerations may be an important factor in the selection of variables. Subsequent sections show the difficulties of obtaining data on supply and the need for proxy measures of the demand for property.

A particular requirement for a forecasting model is *practicability*, that is, it can be easily used for forecasting. One way to ensure this is to have the independent variables lagged (see below). Thus, if the independent variables are lagged by one year, current values can be used to produce a one-year forecast of the dependent variable. However, in annual data, the lag is unlikely to be more than two years. More usually, the requirement is that the independent variables should be forecastable with some reasonable degree of expected accuracy. If this is not the case, what may be an interesting historical model is of little value for forecasting. An example of such a model is discussed in Section 6.5.

Preliminary data analysis and preparation

Having identified appropriate explanatory variables and data sets that may be used to measure or proxy these concepts, it is next necessary to process the data. It is a useful preliminary step to plot the data in time series.[1] The primary purpose at this stage is to check for any obvious errors in the data. These may have arisen at source or in the process of data entry. The plots also help later to identify possible relationships. Bivariate plots of the dependent variable against each of the independent variables also help to identify relationships and can be used to determine if the relationships are linear or if a transformation is required (see below). Thereafter, data analysis and preparation involve deflating the series, tests for stationarity and logarithmic transformations.

It is normal practice to specify a property model in *real* terms, that is, with the effect of inflation removed. Some series are published in real terms but others require to be deflated. This means, for example, that changes in *real* economic variables are used to explain changes in *real* rents. If this is not done, a spurious relationship may be estimated as a result of having inflation in both the dependent and one or more of the independent variables.

An important requirement of modern econometric analysis is that the data series used in the regression estimations should be *stationary*. A detailed discussion of stationarity is beyond the scope of this book, but, in simple terms, stationarity means that the series is not trending upwards or downwards. If the series used in a regression are not stationary, any regression is likely to produce spurious results which arise from the trends rather than the changes in the underlying series.

A series is termed $I(0)$ if it is stationary. If the original series is non-stationary and its first difference is stationary, it is termed $I(1)$; if a series becomes stationary after being differenced k times, it is termed $I(k)$.[2] It is important to test for stationarity prior to any analyses being undertaken and most econometrics software now provides easy-to-use tests.

In practice, many economics series, when converted to logs, are $I(1)$. The levels of (real) series, such as output and income, grow over time and the *amount* by which they grow in each period also grows. However, the growth *rate* is typically stationary. If this is the case, then the first log difference ($\Delta \log(x_t)$) of such a series is stationary, as:

$$\Delta \log(x_t) = \log(x_t) - \log(x_{t-1})$$

$$= \log\left(\frac{x_t}{x_{t-1}}\right)$$

$$= \log\left[\frac{x_{t-1} + (x_t - x_{t-1})}{x_{t-1}}\right]$$

$$= \log(1 + g_t) \approx g_t \qquad \qquad \text{◄Equation 6.1}$$

where: x_t is the value of the real series at the end of period t
g_t is the growth rate during period t.

Accordingly, it is common practice to transform (real) levels data into logs.

There are two other reasons for log transformations. First, the transformation may have the effect of normalising non-normal residuals in a regression thereby ensuring that a model conforms to the normality condition (see below). Second, a log transform allows the regression coefficients to be interpreted as elasticities which accords well with economic theory. For example, a common formulation in economics for the relationship between variables is:

$$R = AD^{\beta}S^{\gamma} \qquad \qquad \text{◄Equation 6.2}$$

where: R, D and S are variables

A, β and γ are constants

β and γ are elasticities.

Taking the logs of both sides of Equation 6.2 gives:

$$\begin{aligned}
\log R &= \log AD^\beta S^\gamma \\
&= \log A + \log D^\beta + \log S^\gamma \\
&= \alpha + \beta \log D + \gamma \log S
\end{aligned}$$

◀Equation 6.3

where $\alpha = \log A$, so $A = e^\alpha$. Equation 6.3 can be estimated using linear regression.

Before beginning model specification, it is normal to calculate summary statistics of the series. These include mean, standard deviation and serial correlation (see below).

Model specification

Having identified the appropriate variables and undertaken the necessary stationarity tests and data transformations, the next stage is to specify a model. In most cases, not only the contemporaneous (x_t) variables but also their lags (x_{t-1}, x_{t-2}, ...) should be included. This allows dynamic effects to be included in the model. It often takes some time for the change in the economy to work through to the property market. For example, an increase in the demand for space will normally take one to three years to be met by completed new construction. Further, if too much new space is provided, there must be an adjustment or correction process while the system adjusts towards an equilibrium. The econometric analysis of such processes is beyond the scope of this book, but, where appropriate, the processes will be mentioned in passing.

There are two general approaches to building a model. The first involves starting with a *broad* possible specification, including all the possible explanatory variables and the maximum number of lags for each which are sensible or which the data permit. Variables, either contemporaneous or lagged, are then eliminated in turn until a final model is produced. The second is to start with a *narrow* specification, perhaps only one explanatory variable, and to build up a model by introducing new variables (including lags of variables already in the model) from a predetermined list. The order of entry is determined by the contribution to the explanatory power of the equation.

In either case, the objective is to find the 'best' specification of the model. However, there are three sets of criteria in defining best. These are: statistical criteria which an acceptable model *must* meet; other statistical criteria which offer guidance (with established rules) for variable selection (see below); and economic theory. The last two of these may offer contradictory guidance.

Statistical procedures are available on most statistical packages which allow the best statistical model to be selected either using a broad approach or a narrow approach. In the former case, at each stage the least important variable is deleted from the model until all of those remaining make a significant contribution to explaining variations in the dependent variable. In the latter, at each stage, a new variable is introduced which contributes most to explaining changes in the dependent variable. This procedure continues until no other variable that makes a significant contribution can be added.[3] Such output still has to be checked for its acceptability under the first set of criteria.

Although these are useful approaches to establish the best statistical model, model building, whether broad or narrow, more usually combines statistics, economic theory and judgement. As discussed above, variables should not be included in the analysis if there is no theoretical justification. However, theory may suggest the primary importance of a variable which fails to pass a standard criterion for its inclusion in the model. In such circumstances, particularly if there are suspicions about the quality of the data, some element of judgement may be required as to whether to include the variable. Further, there may be clear guidance from theory as to whether the effect of a particular variable on the dependent variable is positive or negative. If the best statistical model indicates a contrary relationship, it is not a theoretically acceptable model. However, theory does not always offer guidance on the magnitude of coefficients, and rarely offers clear guidance on the most appropriate lag structure in a model. The latter is usually best determined by the statistics. In practice, in the property market, both theory and data limit the scope of a broad approach.

In applying judgement for initial selection of variables, it is often helpful to plot the dependent variable against each of the possible explanatory variables. This helps to identify the more obvious relationships in the data. For most property variables it is likely that the list of possible explanatory variables will be small, usually less than five. Lags may also be limited: most effects are likely to work through the market within a year, perhaps two at the most in the case of the construction period for new space.

Nonetheless, even parsimonious model specifications may be constrained by data shortage. The short time series available for most property data restricts the number of explanatory variables, particularly when lags are used in quarterly analyses. UK rent data, for example, are available annually from 1971 and quarterly data from quarter three of 1977, giving, at the time of writing, 27 annual and 83 quarterly observations. The manual for one of the better statistical packages states that:

> 'Most authors recommend that one should have at least 10 to 20 times as many observations (cases, respondents) as one has variables, otherwise the estimates of the regression line are probably very unstable and unlikely to replicate if one were to do the study over.' (StatSoft, 1996)[4]

At the other extreme, it is possible (but extremely undesirable) to estimate a regression if the number of observations is greater that the number of independent variables (see below) plus one. Somewhere between the ideal and possible lies a sensible approach to property analysis.

Model estimation

A model is estimated using the statistical technique of linear regression. In the case of two explanatory variables, the model is:[5]

$$y = \alpha + \beta x_1 + \gamma x_2$$

◄Equation 6.4

For any individual observation of y^6:

$$y_t = \alpha + \beta x_{1t} + \gamma x_{2t} + e_t \qquad \qquad \blacktriangleleft \textbf{Equation 6.5}$$

where: y_t is the value of the dependent variable y at time t
$\quad$ x_{1t} and x_{2t} are the values of explanatory (or independent) variables x_1 and
$\quad$ x_2 at time t
$\quad$ α, β and γ are constants
$\quad$ e_t is a stochastic or random error term.[7]

A sample of historical data for y, x_1 and x_2 is used to estimate the model, that is, the values of the parameters α, β and γ. The regression analysis produces the 'best fit' relationship rather than a perfect fit. Thus, at any time, t:

$$y_t = a + b x_{1t} + c x_{2t} + r_t \qquad \qquad \blacktriangleleft \textbf{Equation 6.6}$$

where: a, b and c are the 'best fit' estimates of α, β and γ
$\quad$ r_t is the residual (the difference between the actual value of y and the
$\quad$ value fitted by the regression line) and is analogous to e_t for which it is
$\quad$ an estimate.

The values of a, b and c are selected so that the residuals have a mean of zero and the sum of the squared residuals, Σr_t^2, is minimised. Thus the estimated model explains the maximum possible proportion of the variation in y. The larger the variance of the residuals, the greater the probability that the actual values of the dependent variable are further from the values expected from the model.

The first part, $y_t = a + b x_{1t} + c x_{2t}$, is the *deterministic* part, in other words, the explained variation in y; the second part, r_t, the *probabilistic* part, is the unexplained variation. The larger is the explained variation of y, the better is the estimated model.

Model testing

An estimated model must be tested for conformity to the assumptions made to undertake the regression and also for its 'fit', that is, its quality as a statistical model. It is also important to check that the coefficient signs and values conform to theory. In practice, estimation and testing are undertaken iteratively until a good model is produced.

Four key assumptions are required if the estimated time series regression model is to be valid. These are: linearity; normality of the residuals; constant variance of the residuals; and absence of serial correlation in the residuals. There are formal statistical tests for each of these.

The *linearity* assumption refers to the coefficients. Thus, it is perfectly possible to estimate a model such as:

$$y_t = \alpha + \beta x_{1t} + \gamma x_{2t}^2 + e_t \qquad \qquad \blacktriangleleft \textbf{Equation 6.7}$$

if the inputs are y_t, x_{1t} and x_{2t}^2. As shown in Equation 6.2 above, it is also possible to transform relationships into a linear form which can be estimated:

$$R = AD^\beta S^\gamma$$

As a preliminary to estimation it is good practice to look at a bivariate scatterplot of the dependent variable against each of the independent variables (see above). If the plot makes evident a non-linear relationship, transformation of the variables should be considered. Log transforms, as discussed above, are the most common, but other forms exist.

The *second* assumption is that the errors are *normally distributed* with zero mean. The mean of the residuals is zero by construction. There is a formal statistical test for normality in the residuals, but it also is possible to plot the residuals (both as a time series and as a probability distribution) to identify where any problem may arise.

The *third* assumption is that the variance of the errors is constant. This is known as *homoscedasticity*. Thus, the variance may not increase over time or as the dependent variable increases in size. As above, in addition to the standard statistical test, it is good practice to plot the residuals. Evidence of the variance of the residuals increasing with values of the dependent variable may be evidence of a non-linear relationship and may be overcome by a log transformation.

When examining the plot of the residuals, it may become clear that one (or a sequence) of the observations of the dependent variable does not fit the model, while all the others do. This may be because one factor not in the model was important for those observations but not for others. An obvious example when modelling economic data would be a major strike which lasted long enough or was extensive enough to affect output significantly. An alternative explanation may be a measurement error in the data. In either case, it is appropriate to estimate the model without these observations. This is done by introducing what is known as a *dummy variable*, so called because it is not a true variable.[8] In this case, its effect is to remove the outlying point from the estimation of the model.

The *fourth* assumption is that there is no serial correlation in the errors, that is:

$$\text{cor}(e_i, e_j) = 0 \quad \text{for } i \neq j \qquad\qquad \blacktriangleleft \textbf{Equation 6.8}$$

where: cor is the correlation coefficient and e_i
　　　e_j are errors at times i and j.

If this assumption does not hold, past values contain information about subsequent values.

The other statistical tests consider if the model fits the data well and the importance of each of the variables. They are *R*-squared (R^2); the *F*-value; and *t*-values. *R*-squared is a simple summary figure of the proportion of the total variation of the dependent variable attributable to the independent variables in the model. The closer the value to 1, the greater the proportion of the variation explained by the model; the closer to zero, the less the proportion explained. As *R*-squared increases with the number of independent variables, it is normal to adjust it to take account of the number of variables to produce the adjusted *R*-squared.

The F- and t-values are used to calculate the probability that the regression results have happened by chance and do not represent an actual relationship. The F-value of the regression considers the regression as a whole; and t-values are used to decide whether or not to include individual variables in the model. The lower the significance levels associated with the values, the less the probability of the relationship having occurred by chance. A significance level of 5% for the t-value is usually taken as a cut-off point: above that variables are excluded from the regression. Their use is best illustrated by the example of retail rents in Section 6.3.

Forecasting

When used for forecasting, it is necessary to assume that the estimated relationships will hold in the future, that is, α, β and γ (estimated by a, b and c) will have the same values. In other words, it is assumed that *the future will be like the past*. Known or forecast values of the independent variables are used in the model to produce forecasts of the dependent variable.

One important way to test a model's forecasting ability is to re-estimate it for a restricted time period, say the first three-quarters of the data available, and to use known values of the independent variables to produce *in sample forecasts* of the dependent variable. These can be compared with the actual values in a test of predictive failure.

The procedure outlined above is now illustrated using the example of retail rents in the UK.

6.3 National level forecasts

This section considers model frameworks for rent, supply and demand but focuses on rents. Most of the published work in the area has been on the office sector. Two forms of single equation models for rents are considered. The first, a reduced form model linking rents to demand and supply, is illustrated by the UK retail sector as an example. The second is a simpler model linking rents to vacancies. Finally, the single equation model is developed into a three-equation system with separate equations for supply and demand.

Rent

Rent is determined in the occupier market (see Ball *et al.*, 1998) and is the price paid by a tenant to occupy space. Basic economics suggests that price is determined by the interaction of demand and supply. Thus, a simple model of rents should link rent to demand and supply:

$$R = f(D, S) \qquad\qquad \blacktriangleleft \text{Equation 6.9}$$

where: R is rent
D is demand
S is supply.

This is known as a *reduced form* equation.

Table 6.1▶ UK rent indices

Producer	Frequency	Start date	Disaggregations	Contents
IPD	Annual	1976	Three sectors	Actual portfolio
		1981	Three sectors; standard regions; different subdivisions of London for each sector	
IPD	Monthly	January 1987	Three sectors	Actual portfolio
JLW	Annual	June 1967	Three sectors	Actual portfolio
	Quarterly	June 1977		
HP	Annual	May 1972	Three sectors; standard regions; different subdivisions of London for each sector	Hypothetical properties let at open market rents
	Twice-yearly	May 1977		
	Quarterly	May 1990		
RE	Annual	January 1979	Three sectors	Actual portfolio
	Monthly	January 1987		

Price, demand and supply are all economic concepts and empirical measures of them are required in order to estimate a model. At the UK level, rent has no practical meaning unless it is measured by an index, that is, a weighted average of rents in different locations. This is similar to the Retail Price Index (RPI) which is used in the UK to measure price inflation. One issue worthy of mention when considering rent indices is the rent data included. Some parts of a shop are more valuable than others. In general, within any shop unit, the closer to the street frontage and the closer to ground level, the more valuable is the space and the higher is the rent paid. Rent data are typically for the prime space on the ground floor frontage but may also be averages for all space. A further issue which applies to all property types, but particularly to offices, is that the rent data should be for *effective* rents – that is, with the effects of incentives such as fitting costs and rent-free periods stripped out (see Hendershott, 1995).

A number of rent indices are available in the UK and the main ones are shown in Table 6.1. The selection of a suitable index involves trade-offs among the length of the time series, the frequency, the disaggregations available and the robustness of the construction. For example, if it is intended to produce regional as well as national sector forecasts, the choice is between the Investment Property Databank (IPD) and the Investors Chronicle Hillier Parker (ICHP) indices. The IPD index covers actual portfolios and is now the market benchmark, whereas the HP index is of hypothetical buildings but, as bi-annual data are available, offers twice as many observations for modelling purposes. The other indices in Table 6.1 are the Jones Lang Wooton (JLW) and the Richard Ellis (RE).

Differences in the construction of the series mean that they produce different values. Figure 6.1 shows annual nominal retail rental growth from the IPD, JLW and HP indices. While they all follow the same broad pattern, particularly in recent years, there are differences which would result in different rent models from the same explanatory variables. This illustrates the problems involved in trying to find suitable measures for theoretical concepts.

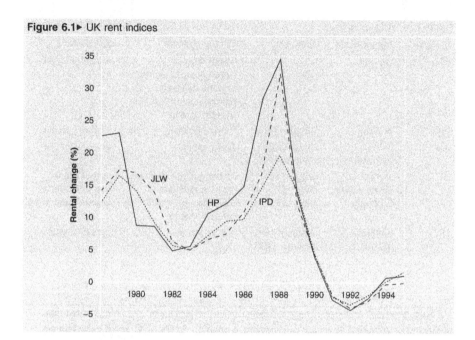

Figure 6.1► UK rent indices

As the demand for property is a derived demand, it has no direct measure. It is a function of the profitability of the occupier. Other things being equal, the greater the profits, the higher the price (rent) which could be paid to occupy the property. Profits depend on the level of activity – in the case of the retail sector this is sales – and the efficiency of the use of inputs in the form of space, labour and capital. On average and over time, variations in sales are likely to be the most important factor in explaining variations in rent. However, increased turnover to floor space ratios through, for example, more efficient stock control, are also factors. The volume of sales can be measured by retail sales and the data are available at the national and regional levels. Data for other plausible demand factors, such as turnover to floor space ratios and retail profits are not available.

The final concept in the proposed model is supply. As supply of retail space increases, other things being equal, price (rent) should fall. Although the stock of space can be measured directly, there are special problems. In the UK, supply data are generally of poor quality and the published data provide an incomplete time series and are for England and Wales only. They are based on data collected for local property taxes and cover all retail space, including secondary and tertiary space.

Total floor space is not a perfect measure of supply as the ratio of sales space to total space is likely to have increased over time. The development of retail chains and pressures to use space efficiently have led to storage areas being converted to sales space, and centralised warehouses and distribution networks providing stock to a shop as it is needed.

Data are also available by sector for new construction orders and for new construction output. These measure the addition to the stock of space but are subject to a number of problems (see Ball *et al.*, 1998: 182–4).

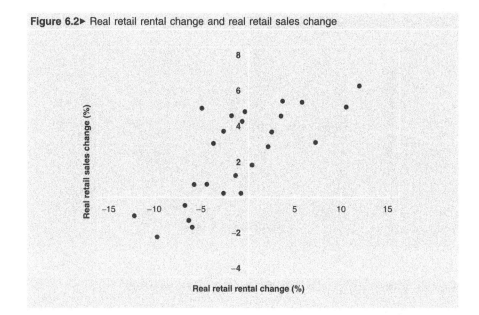

Figure 6.2► Real retail rental change and real retail sales change

The poor quality of supply data is a standard problem in UK property research. Fortunately, this is less of a problem at the national level because supply increases steadily, and the series is not very volatile. In contrast, the lack of reasonable quality supply data creates substantial problems for local market modelling (see below).

One further problem worthy of mention is the mismatch between the properties to which the data sources apply. The supply data contain all shops, the retail sales are a sample of all shops, and the rent data vary according to the index. Some indices are of prime property, others are of general institutional quality property, but none is of all shops.

Tables 6.2 to 6.4 present the results of estimations to derive a suitable model of UK retail rents using retail sales as the demand proxy, and the stock of retail space as the supply variable. All four models use the IPD rental data series. It should be stressed that these are not proposed as good models; rather they are simple estimations which illustrate some of the output from an analysis and also what can go wrong in model building. In all three cases, the models presented are in real terms and in log differences (approximately growth rates) and the variables are stationary.

In Figures 6.2 to 6.5, real retail rental change is graphed, respectively, against real change in retail sales, change in retail stock and one period lagged change in retail stock. From these graphs it is clear that rental change is strongly correlated (0.74) to change in sales, lowly correlated (−0.27) with change in stock but more strongly correlated (−0.58) if the lagged change in stock is considered.

The three main series are graphed in Figure 6.2 as a time series. Note that: rental change is most volatile; change in sales and change in rents follow a broadly similar time pattern; change in supply has low volatility; and there is no clear relationship between rents and supply.

Figure 6.3► Real retail rental change and change in retail stock

Figure 6.4► Real retail rental change and lagged change in retail stock

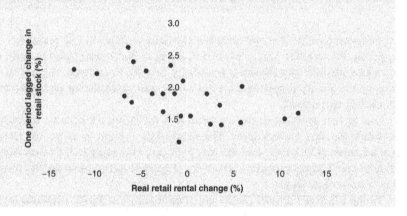

The key points of Model 1 in Table 6.2 are:

1. The constant is not significantly different from zero: the t-value is -0.32 (the rule of thumb is greater than 2 in absolute value for significance) and the significance level is 75.1% (less than 5% is the convention).
2. The coefficients for demand and supply are correctly signed (positive and negative, respectively) and are not statistically significantly different from each other (the null hypothesis that the difference is zero could not be rejected). Taken together with point 1 above, this means that if demand and supply increase at the same rate, rents do not change.
3. The coefficient for supply is not significantly different from zero: the t-value is -0.83 and the significance level is 41.3%. This is a standard problem in such models.
4. The adjusted R^2 is 53.6%, which is reasonable for an estimation in differences, and the F-value for the regression as a whole is significant at 0.0%.

Figure 6.5▶ Real retail rental change, real retail sales change and change in retail stock

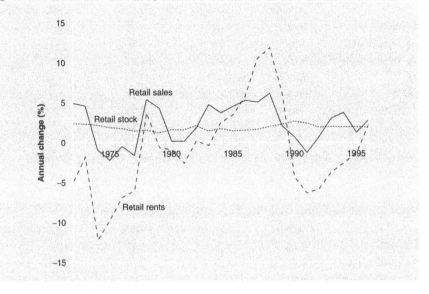

Table 6.2▶ A basic differences model (Model 1 – 25 observations: 1972–96)

	Coefficient	Standard error	t-value	Significance level
Constant	−0.01	0.05	−0.32	75.1%
DL (retail sales)	1.69	0.33	5.08	0.0%
DL (retail supply)	−1.92	2.30	−0.83	41.3%
R^2	57.5%	Diagnostics:		
R-bar^2	53.6%	Serial correlation	1.5%	
F significance level	0.0%	Linearity	82.5%	
		Normality	84.9%	
		Homoscedasticity	32.8%	

Note: DL is difference in the log level, so is approximately the growth rate.

5. The equation passes three of the diagnostic tests but is unsatisfactory because there is evidence of serial correlation in the residuals (values of below 5% indicate a problem).[9]

Model 2 in Table 6.3 uses the lagged value of the change in the log of supply rather than the contemporaneous value. In essence, it is hypothesised that, while demand information becomes widely known quickly, supply information takes some time to filter through to occupiers. As retailers know directly about their sales but find out indirectly about supply, this is not implausible. There are two important differences from Model 1: the supply variable is now significant and has a coefficient of −4.9. This leads to the implausible interpretation that, if supply and demand were to increase at the same rate, rents would fall. Model 2 also suffers from serial correlation.

Table 6.3▶ A lagged model (Model 2 – 24 observations: 1973–96)

	Coefficient	Standard error	t-value	Significance level
Constant	0.05	0.04	1.11	27.9%
DL (retail sales)	1.60	0.30	5.36	0.0%
DL (retail supply)(−1)	−4.90	2.02	−2.42	2.5%
R^2	72.5%	Diagnostics:		
R-bar^2	69.9%	Serial correlation	0.9%	
F significance level	0.0%	Linearity	52.9%	
		Normality	47.3%	
		Homoscedasticity	97.2%	

Note: DL is difference in the log level, so is approximately the growth rate; (−1) denotes a one-period lag.

Table 6.4▶ An adjustment model (Model 3 – 24 observations: 1973–96)

	Coefficient	Standard error	t-value	Significance level
Constant	0.55	0.27	2.07	5.3%
DL (retail sales)	1.47	0.22	6.78	0.0%
DL (retail supply)(−1)	−3.80	2.52	−1.51	14.9%
L (rent)(−1)	−0.40	0.14	−2.95	0.9%
L (retail sales)(−1)	1.14	0.28	4.03	0.1%
L (retail supply)(−2)	−1.45	0.39	−3.76	0.1%
R^2	88.5%	Diagnostics:		
R-bar^2	85.3%	Serial correlation	35.1%	
F significance level	0.0%	Linearity	78.0%	
		Normality	79.1%	
		Homoscedasticity	55.3%	

Note: DL is difference in the log level, so is approximately the growth rate; L is the log level; (−1) denotes a one-period lag; (−2) denotes a two-period lag.

In Table 6.4 lagged levels of all the variables are added to Model 2. This introduces some dynamics into the model and is a variant of what is known as an error correction mechanism (ECM). The basic principle is that if the *level* of rent was different in previous periods (in this simple case, in the previous period) from what it should have been based on the level of demand and supply, then there is a 'correction' or adjustment next period which affects the growth rate.[10] The results of the estimation show an improved model without problems of serial correlation, although the change in supply is significant only at 15%. Applying the same procedure to Model 1 produces the same basic model but a wrongly signed (positive) and insignificant coefficient on the change in supply.

While these estimations illustrate some of the principles involved, they also show some of the problems likely to be encountered when building property rent models. Before considering modelling frameworks for demand and supply, an alternative form of rent modelling, known as rental adjustment models, is considered.

Rental adjustment models

Rental adjustment models link the proportional change in rent to the difference between the actual and the natural or equilibrium vacancy rate. The approach is popular in the USA where vacancy rate data are available. In contrast, they are little used in the UK as the data are not generally available. The basic form of the rental adjustment model is:

$$\frac{\Delta RR_t}{RR_{t-1}} = \lambda(VR^n - VR_{t-1})$$ ◄**Equation 6.10**

where: ΔRR_t is the change in real rent during period t
RR_{t-1} is the real rent at the end of period $t - 1$
VR^n is the natural vacancy rate
VR_{t-1} is the vacancy rate at the end of $t - 1$
λ is an adjustment parameter.

Note that, if a regression line is estimated as:

$$\frac{\Delta RR_t}{RR_{t-1}} = \beta_0 + \beta_1 VR_{t-1}$$ ◄**Equation 6.11**

then $\beta_0 = \lambda VR^n$ and $\beta_1 = -\lambda$, so the natural vacancy rate can be derived from the data.

The model was first used in the housing market but has been applied to the office sector. It has been developed to include operating expenses (Shilling *et al.*, 1987) and an equilibrium rent term (Hendershott, 1995) which links the property market to the capital markets.[11] Such an approach is also used in two published models of the City of London office market (see below). For a full discussion see Ball *et al.* (1998).

Supply

When modelling new development, it is possible to consider either development starts (Wheaton *et al.*, 1997) or development completions (Hendershott *et al.*, 1999). Most such models are based on the profitability of development (building takes place when completed values are greater than construction costs) and generally have the following form:

$$NS = f(CV, CC, IR)$$ ◄**Equation 6.12**

where: NS is new supply
CV is capital values of property
CC is construction costs
IR is the interest rate.

In theory, land values should also be included but data rarely exist.

Hendershott *et al.* (1999) propose an alternative form:

$$CD = \lambda(VR^n - VR_{t-n}, RR^n - RR_{t-n})$$ ◄Equation 6.13

where: *CD* is completed supply
RR^n is the equilibrium rent
RR_{t-n} is an appropriate lag.

In this version, developers use the vacancy rate and rents, both in relation to their equilibrium or natural rates, as indicators of development profitability.

Demand

Demand models typically use the *flexible accelerator* principle. The basic accelerator principle states that the level of new investment in capital goods (in this case, property) is proportional to the change in the level of the output of users of the capital goods. For a given level of technology, the capital to output ratio is constant:

$$K_t^* = vY_t$$ ◄Equation 6.14

where: K_t^* is the desired level of capital stock at the end of period *t*
Y_t is the level of output from users of the capital good
v is the capital to output ratio.

If output rises, and technology is fixed, it is necessary to increase the amount of property, so:

$$NK_t = K_t^* - K_{t-1}$$ ◄Equation 6.15

where: NK_t is desired increase in occupied property
K_{t-1} is the level of capital stock (occupied property) in period $t - 1$.

In practice, demand for new property is not so responsive to changes in output. The costs involved in moving to new space and the need to commit to a lease contract are combined with uncertainty about the continuing need for the new space. Thus, occupiers may first use slack space and will then increase their use of current space until they are confident that the new level of demand will be sustained. The flexible accelerator thus allows for a gradual move from the existing level of stock to the desired level:

$$NK_t = z(K_t^* - K_{t-1})$$ ◄Equation 6.16

where: z is an adjustment parameter with a value between 0 and 1. The desired level can be modelled as a function of employment and the real interest rate:

$$K_t^* = f(E, RR)$$ ◄Equation 6.17

New demand can then be modelled by combining Equations 6.16 and 6.17; however, there is no standard detailed modelling approach.

The accelerator approach has also been used to model new development – that is, supply – on the basis that developers respond to the needs of occupiers. In this case, an adjustment process is required not only because of the actions of occupiers (above) but also those of developers who may delay some building until they are confident of future demand when the space will be completed.

Variants of the use of the accelerator model can be found in Barras (1983), Barras and Ferguson (1987a, 1987b), Benjamin *et al.* (1995), Tsolacos (1998), Wheaton and Torto (1990), Wheaton *et al.* (1997) and Hendershott *et al.* (1999).

The last two of these use it for a demand equation which is part of a larger model of the City of London office market. In each, behavioural equations for demand, development (new supply) and rents are linked to exogenous variables (employment, interest rates and constructions costs). There are also a number of identities covering, for example, vacancy, absorption, occupied and total floor space. These papers show what is possible where data are available, but neither approach is without limitations. Although this work represents the 'state of the art' for multi-equation models, the substantive differences in detail within a broadly common approach indicate the lack of definitive theory and the flexibility for modellers to chose their preferred specifications. Ball *et al.* (1998) provide a detailed critique of the models. Attention now turns to regional and local rent modelling.

6.4　Regional and local rent forecasts

In the previous sections, the models discussed were for the national level. These can be used to produce forecasts to compare property with the other main investment assets. Of far more interest to property portfolio managers are forecasts that would allow them to structure their portfolios and to buy and sell individual properties. This section extends discussion of rent modelling to the regional and local levels. As for the national modelling, most of the published work covers the office market. Regional forecasts are a central part of portfolio construction and local forecasts are important in stock selection. To date, with the exception of two models for the City of London (Wheaton *et al.*, 1997; Hendershott *et al.*, 1999), the published models have been single-equation models rather than multi-equation models.

Regional models

The same data sources used for national level forecasting in the UK are generally also available at the regional level, which means that similar modelling strategies may be adopted. Both the IPD and the Hillier Parker rent indices are available (see above) and most of the macro-economic and supply explanatory variables are available from the same sources. The quality of the data is more problematic, partly because of the smaller samples. Forecasts of the regional versions of the macro-economic variables are available and with a little creativity it is possible to produce a forecast of supply (see above). As at the national level, supply growth tends to be stable.

There are three approaches to modelling regional property markets:

- Estimate a separate model for each region.
- Use all the regional data to estimate a common model for all regions. This is known as the *panel* or *pool* approach.
- Group similar regions together and use the panel approach to estimate models for each group.

Each approach has strengths and weaknesses. The first approach allows the different regional markets to be modelled separately. This means that the explanatory variables used can be customised to a specific market. It also means that different speeds of adjustment of the dependent variable to changes in the explanatory variables can be modelled in each region. However, with few exceptions, the same explanatory variables should apply across the regional markets. One obvious exception is the importance of financial services and overseas trade in determining the demand for office space in the City of London. A further problem is that the data are generally less reliable at the regional level (see below) so that any differences in adjustment processes may be simply the product of poor quality data rather than any inherent differences.

Pooling the data has the effect of ensuring that more robust estimates are made of the model parameters but it imposes a common model on all markets. Several separate panels can be regarded as a sensible compromise approach, although it is necessary to have good reason to separate the regions into groups. Key *et al.* (1994), when considering rent models, argue that the effect of the City of London office market on the office markets in the southern regions of the UK (that is, those contiguous with the South East – the region around London) is so strong that a separate panel should be estimated. The UK office market is dominated by London: 27% by value is in the City of London (the 'square mile'); 76% is in London; and nearly 90% is in the South East region. For the retail and industrial sectors, they argue that a single panel is appropriate. The differences between regions has also been highlighted by Gardiner and Henneberry (1988, 1991) who modelled rent.

A number of matters relating to consistency require careful consideration in regional analyses. If separate models, or even more than one panel model, are estimated, care has to be taken to ensure theoretical consistency. As suggested above, it is unlikely in most cases that different explanatory variables, coefficients of different orders of magnitude or different lag structures could be justified. These might suggest weak theory or results which are artefacts of poor quality data. Similarly, it is important that the regional models, whether separate or panels, conform to the model estimated using national data. Regional models must also be consistent in the sense that the weighted average of rent forecasts and the aggregates of demand and supply forecasts must be the same as the national forecasts of the variables.

In general, it is possible to estimate sensible models at the regional level but these have weaker explanatory power. However, it is often difficult to justify (statistically) the inclusion of supply in the rent models. Key *et al.* (1994), for example, are unable to include supply in their single panel of the UK retail markets or their northern office markets panel. They suggest that it is possible that these markets are primarily demand driven and that little speculative development occurs, but they also acknowledge that data problems may also be an explanation. Other authors have encountered similar problems.

One approach beginning to appear in the US literature, but which has not yet been applied in the UK, is spatio-temporal autoregressive (STAR) models. These enable the modelling of changes in one region to be transmitted through time to surrounding regions. This would allow the modelling of two perceived wisdoms of the UK market: that significant changes (booms or declines) are first experienced in London and then 'ripple' through the regional markets; and that spill-over effects of the growth of London have spread to other regions in the south of England (see MacGregor and Schwann, 1999, for a preliminary analysis).

Although administrative or contiguous geographical regions have been much used in the UK for analysis of the property market for forecasting and portfolio construction (see Chapter 9), several authors have raised concerns about their usefulness. The central concern is that the regions cover a wide variety of local markets, ranging from declining conurbations to growing market towns and tourist centres. Regional data, therefore, provide an average of very different local markets which have performed, and can be expected to perform, in different ways. However, attempts to find alternative classifications based on groups of local markets and using socio-economic and property data have not been successful (see below and Chapter 7).

The same problem, but on a greater scale, has been tackled by US authors. There the focus has been on portfolio construction (and hence on returns) rather than on market modelling. In the absence of property market data, regions have been classified using economic data, thus concentrating on demand and excluding differential supply responses from the analysis. The general conclusion of such work is that non-contiguous regions, with similar economic structures, rather than contiguous geographic regions provide most scope for diversification (see, for example, Mueller, 1993).

Local level modelling

In general, the lower the level of aggregation, the greater are the problems encountered in modelling the property market. The local level presents a number of theoretical and practical difficulties (see Ball *et al.*, 1998). The discussion below illustrates some of these issues by reference to the modelling of retail rents. These include:

- ▶ defining the boundaries of local market areas within which to consider the interaction of supply and demand
- ▶ overlapping demand across several local market areas
- ▶ competing supply in nearby areas
- ▶ the availability and quality of the data
- ▶ unlike the national and regional levels, new supply tends to be built infrequently and in large amounts relative to the size of the existing supply
- ▶ the importance of factors which are not easy to incorporate within a model.

In building a model, it is important to define an area of study within which an increase in demand will be met, first, with an increase in rents and then with an increase in supply. Interactions across boundaries complicate the modelling. At the UK level, this is not a significant issue: a state, largely comprising a single large island, is not affected by significant cross-border demand or supply effects in the property market. For other European countries it may be a factor. The UK regions, while partly an administrative convenience, generally have some property market relevance as each is based around one or more significant cities. There are cross-borders effects but, with some exceptions, mainly around London, these are unlikely to be great.

At the local level, the problems may increase. It is relatively easy to define market areas for free-standing towns and cities, and methods have been developed, particularly by retail geographers, for calculating the retail demand for a centre (see Field and MacGregor, 1987, for a discussion). However, in many areas, shoppers have a choice from a number of competing centres. Some shoppers choose one centre while others choose competing centres, so defining market areas is problematic. The solution is to apportion the population according to the size and accessibility of the competing centres.

As the distribution of population changes with new residential developments and as new supply becomes available, the sizes of catchment populations change over time. Thus, what happens in one local market can have a significant effect on adjacent areas. Matters are further complicated by the development of out-of-town centres. Choices between the city centre and out-of-town centres are linked to accessibility and so to the road network.

At the local level, data quality and availability are major problems. A major investor with a UK property portfolio would be interested in perhaps 50 major retail centres. Rent data are publicly available for each of the sectors but are typically annual and only short time series are publicly available. As for all such analyses, the rental data are for prime or institutional property whereas demand and supply measures cover all property.

The demand proxies used at the national and regional levels, such as retail sales, GDP, business services output and retail sales, are not available at the local level. In their absence, other proxies have to be found. It is possible to use population and employment as the explanatory variables: population drives retail sales; and employment is linked to output and so to rents in the office and industrial sectors. Estimates of retail demand based on real incomes within any given drive-time are commercially available. The approach can be elaborated to include affluence variables such as car ownership and social class (see Jackson, 1998). However, there are two fundamental problems. First, the areas for which the data are collected are not the (changing) local property areas of interest to a modeller, but administrative areas. This means that the data have to be apportioned to map, approximately, data collection areas to market areas. Second, although the basic population and employment data are at best annual, the affluence variables are available only from the ten-yearly census.

Supply data are, if anything, more problematic than rent and demand data. There are no consistent data for supply covering the major UK centres. For England, annual data are available for local authority areas from the 1970s but the time series is incomplete. As above, this also presents some problems for matching data collection areas with market areas. Further, the data were collected for local property tax purposes and are for all shops, including secondary and tertiary centres, and not just prime or central area shops. The data have never been available in Scotland. Some indications of new supply can be obtained from planning registers but that would involve substantial costs in collection. Even if this were done, assumptions would be required when, and if, developments with planning permission would be completed, or even begun.

The absence of quality supply data is a particular problem at the local level because, while supply is not particularly important at the national and regional levels, it is crucially important at the local level. New supply at the local level comes infrequently and in large amounts which may add substantially to the existing stock and so can have a dramatic impact on rents. It may also increase the attractiveness of a centre and so attract shoppers from other local areas.

A further problem is that many of the factors which affect rents in a local property market are difficult or impossible to include in a model. Examples are: road building (which affects the accessibility of competing centres and so alters market boundaries); planning policy (which may change from restricting out-of-town developments to permitting them and so change the responsiveness of new supply to rent rises); pedestrianisation (which can affect the accessibility and attractiveness of a centre); car parking availability and costs; the changing mix of shops, including 'magnet' retailers such as major supermarkets or stores; and the availability of complementary facilities, such as cinemas and restaurants. Most require consideration but could not be incorporated into a model.

The rent data available at the local level are of provable rents from rent reviews for existing tenants or from new lettings. Rent reviews are based on market evidence from previous rent reviews or new lettings. If all the market evidence is from reviews rather than new lettings, it is possible that, in a rising market, what is provable could fall below what would be obtainable in an open market letting. During the UK's retail boom of the late 1980s, this became a significant issue. In many towns, surveyors believed that new lettings would set a higher level of rent but could not prove this. In these circumstances, changes in rents could be attributable to either or both changes in the correct rent (based on the interaction of demand and supply) and adjustments to correct rent through open market evidence. At the other extreme, in weak markets, landlords offer incentives such as rent-free periods and fitting costs which, if taken into account, reduce the quoted 'headline' rent. Both circumstances create problems for modelling at the local level which are much less of an issue at more aggregate levels.

The above problems mean that, in most cases, it is not possible to build a meaningful time series model from local data. The exceptions are the office markets of large cities where data may be more readily available and the dominance of the city in the surrounding region means that a number of the conceptual problems do not apply. The works of Wheaton et al. (1997) and Hendershott et al. (1999) are good examples of what can be done, although the problems of data covering inappropriate areas and competing sub-markets are encountered (see Ball et al., 1998, for a detailed critique of these models).

If the problems cannot be overcome, there are a number of other possible approaches. First, the coefficients of national and regional models can be used with local inputs based on possible future scenarios to allow consideration of the possible paths of rental change. Subjective modifications of the forecasts may be required for factors not in the model. The problem is that a region can contain towns of very different sizes and characteristics. For example, in some towns, an increase in demand is likely to result in extra expenditure at a regional centre rather than locally, so there is no reason to suppose that the coefficients will be the same. This emphasises the problems in defining local market areas. Second, the national and regional forecasts can be used to provide a benchmark forecast which is then moderated upwards or downwards depending on the expected balance of demand and supply in the local market. Care is required that forecasts thus produced for towns in a region are consistent with the regional aggregate figure. However, local market forecasts are unlikely to cover the entire region, so the matching cannot be exact. In practice, some combination of the last two approaches may produce sensible results.

A third possibility is to use cross-sectional analysis (or a panel with a small number of time points) for a number of towns to estimate a common model. This overcomes some of the data problems and the use of a limited panel would allow some dynamics to be introduced. This approach has the potential for identifying market areas where rents appear to be too high or too low and so help to assess their likely future movements. Given the problems mentioned above of different types of local markets requiring different models, the estimation of a number of panels for towns of a similar type is a possibility. The problem here is defining 'similar'. A number of authors have noted that the available classifications of towns in the economic geography literature seems to offer no guidance on property market outcomes (see Hamelink *et al.*, 1998; Hoesli, Lizieri and MacGregor, 1997; and Jackson, 1998). One fundamental problem is that the classifications use only demand-related variables and supply responses are omitted from the analyses. These authors have also tried to produce alternative classifications based on local property market data but with limited success. Such classifications as have been suggested tend not to be temporally stable.

A fourth and different approach involves taking a longer term view and to use a pricing model of the form outlined in Chapter 5. Risk premia can be calculated based on the level of diversification in the local economy. Thus, a diversified economy has low risk as the failure of one industry is unlikely to have a serious effect on property demand; and a local market dominated by a single industry is risky. Liquidity in the local market can be obtained from a survey of investors. Long-term rental growth can be based on forecasts of the economic sectors in the local market area. In the absence of contrary information, national figures can be used for depreciation (see Chapter 8). Current capitalisation rates can be obtained from a survey of transactions. The inputs can, thus, be assembled to consider long-term expected returns which can, in turn, be used to assist with portfolio structuring and stock selection. This approach could be used in conjunction with the information provided by the cross-section method on rental levels.

Current rents are determined by the interaction of demand and supply in the occupier market. In contrast, capitalisation rates are determined in the investor market and are driven by expectations and changes in expectations. Total return includes capital return and so is substantially influenced by changes in capitalisation rates. Accordingly, modelling either is much more difficult than modelling rents.

In Chapter 5 it was shown that the property capitalisation rate could be written, approximately, as:

$$k = RF + RP - G + d \qquad \qquad \text{◀Equation 6.18}$$

where: k is the capitalisation rate for a new letting
RF is the risk-free rate determined in the gilt market
RP is the risk premium
G is the expected rental growth rate
d is the expected depreciation rate.

Thus, a change in the capitalisation rate is driven by a change in any of these components. Gilt yields are determined in the bond market but a change should affect the property capitalisation rate through the risk-free rate. The other three inputs are determined in the property market. A purely historical model could be estimated linking the property capitalisation rate to variables which measure or proxy the factors driving yields.

Hetherington (1988) proposes a model with the following explanatory variables:

► the yield on long-dated gilts
► institutional investment in property (smoothed by constructing a moving average)
► bank lending to property companies.

The second and third of these are linked to the risk premium and expected long term growth. Increased institutional investment and bank lending to property companies suggests that expected returns will be high, perhaps as a result of higher expected growth, and that risk will be lower. However, even if these expectations turn out to be incorrect, there is an element of self-fulfilment. Increased demand for property will increase the price and decrease the yield, regardless of changes in investment fundamentals. Key *et al.* (1994) produce a broadly comparable model.[12]

One problem in modelling the capitalisation rate is that the series is very stable and is characterised by slow movement downwards, then upwards, over long periods. In contrast, a number of the possible explanatory variables are more volatile. This is why Hetherington's model requires the smoothing of institutional investment.

While the above approach may result in a plausible model, it is of limited value for forecasting as the explanatory variables are probably more difficult to forecast than the capitalisation rates themselves. Clearly, an alternative approach is required. There are two possibilities. First, the basic framework above can be used, rather more subjectively and with scenarios for the key explanatory variables, to forecast likely directions of movement in the capitalisation rate. This provides useful information for investors, particularly as movement tends to be small and slow.

A second approach is to use the pricing model of Chapter 5. The model will produce a forecast of average returns until the market adjusts to equilibrium after five years. Alternatively, it is possible to assess what the correct yield should be in each future year based on the assumption that the cash flow forecasts are correct. It then becomes necessary to impose a process of market adjustment towards these correct figures. Experience suggests that this may sometimes be slow. These annual yield forecasts can then be combined with the rent forecasts to produce annual return forecasts. Whatever approach is taken, yield forecasts are an essential input to forecasts of returns which are, in turn, an essential input into portfolio construction.

6.6 Summary and conclusions

This chapter has set out the basic framework for the forecasting of property markets. Such forecasts are an important input for explicit cash flow analyses. They have become progressively more common as longer time series of property data have become available and as software has developed.

The basic steps in model building are: conceptualisation of the model; preliminary data analysis and preparation; model specification, estimation and testing; and forecasting. The steps were outlined but it was stressed that the process of estimating a forecasting equation is not linear and that iteration among the steps is essential.

▶ Conceptualisation of the model involves the application of economic theory to identify plausible explanatory variables. A simple model is generally to be preferred as the theory upon which it is based is likely to be more plausible. When selecting explanatory variables it is essential to consider the availability of data to measure the concepts. With plausible lags on some variables of one year, forecasts of these variables are required to generate property market forecasts for more than one year.
▶ The next step is preliminary data analysis and preparation. The data series are graphed as time series and as bivariate plots of the dependent variable against each of the independent variables. This helps identify possible outliers and non-linear relationships. The data are then deflated as necessary to remove inflation, transformed to logs where appropriate and tested for stationarity. For many economic series, first differencing the logarithmic transformation of the original levels variables produces a stationary variable. The log difference is the growth rate of the original series. Log transforms allow coefficients to be interpreted as elasticities and may also help to normalise the data.
▶ Model specification involves consideration of lagged as well as contemporaneous variables. There are two general approaches: a broad approach starts with as

many variables and lags as practicable and proceeds to eliminate insignificant variables; a narrow approach starts with a small number of variables and adds further variables if they are significant. In practice, in the property market, data limitations mean that a broad approach is relatively narrow. Although these approaches can be automated in many statistical packages, a more flexible approach to variable deletion and inclusion which permits the application of theory is preferred.

▶ The estimation is done using linear regression. This makes four main assumptions which must be tested: the absence of serial correlation in the errors, linearity, normality in the errors and homoscedasticity.

▶ Other statistics, such as R^2 and F- and t-values assist in model selection. Throughout, the results must accord with theory. While theory may be clear on the signs of coefficients, it rarely has anything to say about the appropriate lag structure and often has little to say about magnitudes.

▶ Finally, the chosen model may be tested for its in-sample forecasting ability and may then be used to forecast. This assumes that the relationships estimated from historical data will hold in the future. Forecast values of the independent variables are used to produce a forecast of the dependent variable.

Having set out the basic approach, a framework for building a reduced form model for retail rent was considered. This involved supply and demand as independent variables. Rent is measured by an index, demand for retail space is proxied by the volume of retail sales, and supply is measured directly. In the UK, supply data are of poor quality. An alternative rental adjustment model which links rental change to the difference between the natural and actual vacancy rates was also considered. This is widely used in the USA where vacancy data are more readily available and has been used in the City of London office market.

Rather than have supply and demand determined exogenously, it is possible to have them determined by separate equations with a model framework. New supply (development) is generally modelled using variables which are linked to development profitability or a flexible accelerator approach. Demand modelling is more complex and combines a flexible accelerator approach with desired space as a function of occupier output and rent. Two published models of the City of London office market use employment, interest rates and construction costs as exogenous variables. Demand, supply and rent are determined together within the model framework.

The basic framework can be applied to regional models. Three approaches are used: (1) model each region separately; (2) combine them all in a single panel; and (3) use several panels which combine similar regions. Two forms of consistency are required: theoretical and numerical. The separate models must accord to the same economic theory and separate rent forecasts must sum to the national model results. There are, however, problems with the region as a unit of analysis as it is an average of possibly very different local markets. Attempts in the UK to find an alternative approach based on combining similar local markets have met with only limited success.

Local modelling suffers from a number of problems:

- defining the boundaries of local market areas within which to consider the interaction of supply and demand
- overlapping demand across several local market areas
- competing supply in nearby areas
- the availability and quality of the data
- unlike the national and regional levels, new supply tends to be built infrequently and in large amounts relative to the size of the existing supply
- the importance of factors which are not easy to incorporate within a model.

Models of yields are based on variables which are linked to the components of a yield: the risk-free rate, the risk premium and growth. The variables included are: the yield on long-dated gilts; institutional investment in property (smoothed by constructing a moving average); and bank lending to property companies. The second and third of these are linked to the risk premium and expected long-term growth. Such models, while interesting in exploring the factors which drive yield change, have little value in forecasting because the independent variables are so difficult to forecast. An alternative approach involves the use of the pricing models (derived in Chapter 5) to determine what the correct yield should be in each of the next periods based on forecasts of income. An adjustment process to the correct yield may then be imposed to derive period-by-period returns.

Notes

1. There is no hard and fast rule on the precise order of these preliminaries. It may be more appropriate to graph the data prior to transformation; and it may be appropriate to test for stationarity before considering a log transformation.
2. Differencing once simply means taking the current value of the series from the previous value to produce a new series. Differencing twice means repeating the procedure on the series produced from first differencing.
3. Forward Stepwise Regression starts with one independent variable. Others are individually added or deleted (if they are already in the model) at each step of the regression according to predetermined criteria for determining their importance in the model. This proceeds until the 'best' model is obtained. In Backward Stepwise Regression, the independent variables are removed from the regression equation one at a time until the 'best' model is obtained.
4. There are tests of coefficient stability, but these are not considered here.
5. This may be considered as an 'average' relationship for all values of y, x_1 and x_2 and is analogous to taking the expectations of the variables. As the expectation of the error term is zero, it does not appear. It contrasts with Equation 6.5, which refers to the individual values of the variables.
6. See, for example, Gujarati (1992) for a fuller discussion. Note, however, that while some authors use u to refer to the error and e to refer to the residual, others use e for the error. As this book is not aimed at specialist econometricians, the more straightforward use of e for the error and r for the residual is adopted.
7. This can be thought of as the effect of a large number of relatively unimportant variables which are not included in the model.
8. Dummy variables take the values of 1 and 0, depending on whether the observation is of interest. Thus, in the example of a strike, the dummy variable would have a value of

1 during the strike and 0 at all other times. Dummy variables have a variety of other uses, including estimation of seasonal effects in a time series, but these are rarely of importance in property market analysis and are not covered here.

9. It is possible to use methods of estimation other than ordinary least squares (the standard regression procedure) to model the serially correlated error, but these are beyond the scope of this text.

10. If, in levels, y, x_1 and x_2 are co-integrated, $y_t = \beta_0 + \beta_1 x_{1t} + \beta_2 x_{2t} + e_t$ and if β_0, β_1 and β_2 are estimated by b_0, b_1 and b_2, then $y_t - b_0 - b_1 x_{1t} - b_1 x_{2t}$ (the residuals) is a stationary series with zero mean. The lagged levels can be included in the estimation and the error correction mechanism coefficients derived.

11. The equilibrium rent is derived from a user cost of capital model. Thus, the rent rate must be competitive with the returns on other investments. It contains a risk-free rate of return derived from the government bond market.

12. These and many similar studies were produced before tests for stationarity and co-integration were a standard part of applied econometric analysis in the property market.

Further reading

Antwi, A. and Henneberry, J. (1995) Developers, non-linearity and asymmetry in the development cycle, *Journal of Property Research*, 12(3), 217–39.

Ball, M., Lizieri, C. and MacGregor, B. D. (1998) *The economics of commercial property markets*, Routledge, London.

Barras, R. (1983) A simple theoretical model of the office development cycle, *Environment and Planning A*, 15, 1361–94.

Benjamin, J. D., Judd, D. and Winkler, D. T. (1995) An analysis of shopping center investment, *The Journal of Real Estate Finance and Economics*, 10(2), 161–8.

Benjamin, J. D., Judd, D. and Winkler, D. T. (1998) A simultaneous model and empirical test of the demand and supply of retail space, *The Journal of Real Estate Research*, 16(1), 1–14.

DiPasquale, D. and Wheaton, W. C. (1996) *Urban economics and real estate markets*, Prentice Hall, Englewood Cliffs (NJ).

Gardiner, C. and Henneberry, J. (1988) The development of a simple regional office rent prediction model, *Journal of Valuation*, 7, 36–52.

Gardiner, C. and Henneberry, J. (1991) Predicting regional office rents using habit-persistence theories, *Journal of Property Valuation and Investment*, 9, 215–26.

Giussani, B., Hsia, M. and Tsolacos, S. (1993) A comparative analysis of the major determinants of office rental values in Europe, *Journal of Property Valuation and Investment*, 11, 157–72.

Harris, R. (1995) *Using co-integration analysis in econometric modelling*, Prentice Hall/Harvester Wheatsheaf, Hemel Hempstead.

Hendershott, P. (1995) Real effective rent determination: evidence from the Sydney office market. *Journal of Property Research*, 12(2), 127–35.

Hendershott, P., Lizieri, C. and Matysiak, G. A. (1999) The workings of the London office market, *Real Estate Economics*, 27(2), 365–87.

Key, T., MacGregor, B. D., Nanthakumaran, N. and Zarkesh, F. (1994) Economic cycles and property cycles, Main report for Understanding the property cycle, Royal Institution of Chartered Surveyors, London.

Maddala, G. (1994) *Econometrics*, McGraw-Hill, Maidenhead.

McGough T. and Tsolacos, S. (1995) Forecasting commercial rental values using ARIMA models, *Journal of Property Valuation and Investment*, 13(5), 6–22.

McNamara, P. F. (1991) The problems of forecasting rental growth at the local level, in *Investment, procurement and performance in construction*, Venmore-Rowland, P., Mole, T. and Brandon, P. (Eds), E & FN Spon, London, 64–76.

Mills, T. C. (1991) *Time series techniques for economists*, Cambridge University Press, Cambridge.

Morrison, N. (1997) A critique of a local property forecasting model, *Journal of Property Research*, 14(3), 237–55.

Orr, A. M. (1996) Modelling regional industrial property markets: equilibrium or disequilibrium?, *Aberdeen Papers in Land Economy*, 96–10, University of Aberdeen.

Shilling, J. D., Sirmans, C. F. and Corgel, J. B. (1987) Price adjustment process for rental office space, *Journal of Urban Economics*, 22, 90–100.

Shilling, J. D., Sirmans, C. F. and Corgel, J. B. (1992) Natural office vacancy rates: some additional estimates, *Journal of Urban Economics*, 31, 140–3.

Tsolacos, S. (1998) Econometric modelling and forecasting of new retail development, *Journal of Property Research*, 15(4), 265–84.

Tsolacos, S., Keogh, G. and McGough, T. (1997) Modelling use, investment and development in the British office market, *Aberdeen Papers in Land Economy*, 97–01, University of Aberdeen.

Wheaton, W. C. (1987) The cyclic behavior of the national office market, *Journal of the American Real Estate and Urban Economics Association*, 15(4), 281–99.

Wheaton, W. C. and Torto, R. G. (1990) An investment model of the demand and supply for industrial real estate, *Journal of the American Real Estate and Urban Economics Association*, 18(4), 530–47.

Wheaton, W. C., Torto, R. G. and Evans, P. (1997) The cyclic behaviour of the Greater London office market, *The Journal of Real Estate Economics and Finance*, 15(1), 77–92.

Chapter 7

Risk and portfolio theory

7.1 Introduction

The basic principles of risk were considered in Chapter 3 where the discussion was mainly of single asset risk. Chapter 5 considered pricing models and identified income growth, risk and depreciation as key inputs. Income forecasting was covered in Chapter 6 and depreciation is considered in Chapter 8. This chapter considers risk in more detail. Two broad aspects are covered: first, Modern Portfolio Theory (MPT), which considers the efficient trade-off between expected return and risk in a portfolio context, that is, when individual assets are combined; and, second, the Capital Asset Pricing Model (CAPM), which deals with the appropriate risk premium for an asset's risk.

The future return on an investment is unknown but it is possible, in theory, to estimate the probability distribution of future returns. The *mean* or *expectation* of the distribution is the 'best guess' for the future return; the standard deviation of the distribution is the conventional measure of risk and shows how widely spread expected returns are around the expectation. In practice, the standard deviation of *historical* returns is typically used as a convenient proxy.

When assets are combined in a portfolio, the expected return (the expectation) of the portfolio (the combination of the individual assets) is a simple weighted average; but risk is more complex and involves correlations between the assets. Combining assets with low correlations provides opportunities to diversify risk. Thus, an understanding of portfolio risk is essential if portfolios are to be constructed efficiently.

Section 7.2 considers Modern Portfolio Theory, a theoretical framework for calculating risk and return when assets are combined in a portfolio. MPT identifies all possible combinations of risk and return and excludes those which are sub-optimal. An investor may choose from the resultant set of *efficient portfolios* depending on their risk/return trade-off. MPT is one of the methods discussed in Chapter 10 to determine the role of property in multi-asset portfolios and in Chapter 12 to consider international property investment.

An extension of the MPT approach is considered in Section 7.3. The Capital Asset Pricing Model examines risk/return relationships in investments when the market is in equilibrium, that is, when prices are fair. It introduces the notion that the risk of an asset can be separated into two components: a market (systematic) factor and a unique (specific) factor. Under CAPM, investors are rewarded not for total risk but for systematic risk: the specific risks can, in theory, be diversified away. The CAPM enables the appropriate risk premium for an investment to be estimated.

The basic MPT approach is combined with the notion of specific risk from CAPM in Section 7.4 to set out a practical procedure for property portfolio construction. It considers return and risk *relative to a benchmark*. This risk is known as the *tracking error* and is central to most portfolio strategy in practice. It provides the basis for the discussion of practical portfolio construction in Chapter 9.

A further dimension of risk is considered in Section 7.5. For many investors, the risk of not being able to meet their liabilities, such as future insurance payments or pensions, is a central business concern. Risk in this context is not being able to meet liabilities and investment portfolios must be structured to ensure that liabilities can be met. The central concern is the sensitivity of future cash flows to changes in the discount rate and *duration* is its conventional measure. Portfolios are constructed so that the durations of assets and liabilities are matched. Finally, Section 7.6 provides a summary and conclusions.

7.2 Modern Portfolio Theory

Modern Portfolio Theory is based on the trade-off between return and risk in a portfolio context. Most investors hold portfolios of assets rather than a single asset. Therefore, the returns and risks of individual assets are important only because they are used to calculate the portfolio return and risk. Until the 1950s, the notion of diversification was treated subjectively: there was an awareness of the issues but there was no formal analysis. In the 1950s, Markowitz developed a rigorous analytical approach to portfolio diversification (Markowitz, 1952, 1959). It was developed initially for portfolios of shares. The purpose is either:

▸ for a given level of risk, to maximise return; or
▸ for a given return, to minimise risk.

The output of the analysis is the proportion of funds (the weight) to be invested in each asset, and a measure of the expected return and the risk.

The first stage is to compute the expected return and risk of each individual asset and to use these to calculate the portfolio expected return and risk from all possible combinations of weights (w_i).

Combining assets

Chapter 3 set out the basic formulae for return and risk of a portfolio of investments and illustrated these with examples. The expected portfolio return is the weighted average of the expected returns:

$$E(R_P) = \Sigma\, w_i E(R_i)$$

◄Equation 7.1

where: $E(R_P)$ is the expected portfolio return
$\quad\quad\;\; w_i$ is the proportion of the portfolio in asset i
$\quad\quad\;\; E(R_i)$ is the expected return for asset i.

Note that:

$$\Sigma\, w_i = 1$$

◄Equation 7.2

The expected portfolio risk is:

$$\sigma_P = \sqrt{\Sigma\Sigma\, w_i w_j \sigma_i \sigma_j \rho_{ij}}$$

◄Equation 7.3

where: σ_P is the portfolio risk
$\quad\quad\;\; \sigma_i$ and σ_j are individual asset risks
$\quad\quad\;\; \rho_{ij}$ is the correlation coefficient for assets i and j.

For a two-asset portfolio, this can be expressed as:

$$\sigma_P = \sqrt{w_1^2 \sigma_1^2 \rho_{11} + w_2^2 \sigma_2^2 \rho_{22} + w_1 w_2 \sigma_1 \sigma_2 \rho_{12} + w_2 w_1 \sigma_2 \sigma_1 \rho_{21}}$$

◄Equation 7.4

Note that ρ_{ii}, the correlation of an asset with itself, is 1 in all cases, and that $\rho_{ij} = \rho_{ji}$ for all i and j. Thus Equation 7.4 simplifies to:

$$\sigma_P = \sqrt{w_1^2 \sigma_1^2 + w_2^2 \sigma_2^2 + 2 w_1 w_2 \sigma_1 \sigma_2 \rho_{12}}$$

◄Equation 7.5

If assets 1 and 2 are perfectly correlated, then $\rho_{12} = 1$, and Equation 7.5 becomes:

$$\sigma_P = \sqrt{w_1^2 \sigma_1^2 + w_2^2 \sigma_2^2 + 2 w_1 w_2 \sigma_1 \sigma_2}$$
$$= \sqrt{(w_1 \sigma_1 + w_2 \sigma_2)^2}$$
$$= w_1 \sigma_1 + w_2 \sigma_2$$

◄Equation 7.6

More generally, if $\rho_{ij} = 1$ for all i and j:

$$\sigma_P = \Sigma\, w_i \sigma_j$$

◄Equation 7.7

Thus, portfolio risk is at its maximum – the weighted average of individual asset risks – only when all correlations are +1. In practice, no two assets are perfectly positively correlated as their returns are affected by different factors. Thus, the risk is reduced when assets are combined in a portfolio, and the amount by which risk is reduced depends on the correlations between the assets: other things being equal, the lower the correlations, the greater the risk reduction. This is the basis of diversification and portfolio construction.

The minimum theoretical value for portfolio risk is when all the correlations are at their minimum values of −1. For a two-asset portfolio, Equation 7.5 simplifies to:

$$\sigma_P = w_1 \sigma_1 - w_2 \sigma_2$$

◄Equation 7.8

Figure 7.1► The feasible set and the efficient frontier

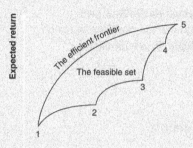

Thus, in theory, for a two-asset portfolio with perfect negative correlation, it is possible to construct a portfolio with no risk ($\sigma_P = 0$) by choosing weights such that $w_1/w_2 = \sigma_2/\sigma_1$.

However, this is a theoretical result with no practical relevance as no two assets have perfect negative correlation.[1]

The efficient frontier

The above formulae can be used to calculate the feasible combinations of risk and return. For a five-asset portfolio, these are illustrated in Figure 7.1 which shows the characteristic umbrella shape of the feasible set. Note that some portfolios are sub-optimal: it is possible to obtain either higher expected return for the same risk, or lower risk for the same expected return. In either case, the optimum point is on the upper boundary. This line is known as the *efficient frontier* and comprises all the optimum combinations of risk and return. Points along the frontier are mean-variance efficient: that is, a higher expected return bears a higher risk (see Chapter 3).

Selecting the investor's optimum portfolio using indifference curves

Having identified all the possible optimum points, it is then necessary to select one of these. As the points are mean-variance efficient, the decision depends on the risk/return trade-off of the investor. This can be determined by their indifference curves. Consider the curve I_1 in Figure 7.2: starting from one point, an investor could move along the curve to a new point with a higher level of risk and a higher level of expected return, and be indifferent between the two points. In other words s/he is rewarded with an appropriate amount of expected return for the additional risk. Similarly, the investor could move to a lower level of risk and a lower expected return, and again be indifferent between the points. The curve maps out combinations of risk and return among which an investor is indifferent. Thus, along this curve, an investor has the expected return s/he requires for the risk taken.

Starting from different points, it is possible to construct a family of indifference curves (such as I_1 to I_4) for any investor. The higher the indifference curve, the

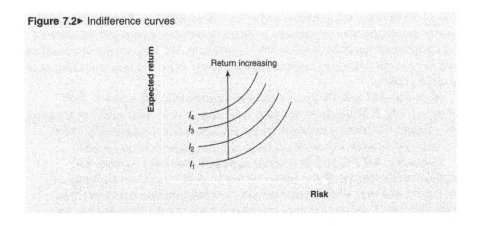

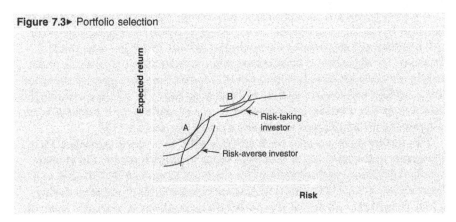

higher the expected return for a given level of risk. Thus, an investor would wish to be on the highest possible indifference curve. The optimum portfolio for an investor is, therefore, the point on the efficient frontier at which s/he is on the highest indifference curve. This is where an indifference curve is tangent to the efficient frontier.

Consider two investors, one risk averse, the other a risk taker. The former requires a higher additional return for each unit of risk, so his/her indifference curves are steeply sloping. The risk-taking investor, on the other hand, has shallow sloping indifference curves. When these are drawn as in Figure 7.3, the result is that the risk-averse investor chooses an optimum portfolio at point A to the left of the efficient frontier, while the risk-taking investor chooses a portfolio at point B to the right. The output of the analysis is the optimum risk and expected return and the weights for the individual assets.

Practical applications

MPT was developed by Markowitz for application to the selection of individual shares. Initially, however, MPT was rarely used in practice as the data requirements are substantial. A portfolio of *n* assets requires the estimation of *n* expected returns, n risk measures and $n(n - 1)/2$ correlations. Thus, a three-asset portfolio requires nine calculations, a 30-asset portfolio requires 495 calculations and a 100-asset portfolio requires 5150 calculations. With the advent of computers which have alleviated the computational problems, MPT has been more and more used in practice.

Historical data provide no easy answer in determining the inputs to MPT: Sharpe (1990: 7–37) suggests that such data 'appear to be quite useful with respect to standard deviations, reasonably useful for correlations and virtually useless for expected returns. For the latter, at least, other approaches are a must.'

In practice, MPT is used to combine aggregate asset classes rather than individual investments. These classes should be chosen to ensure similarity of returns and risks within each class and dissimilarities between classes. The identification of appropriate asset classes is not a trivial problem and has been the subject of much research in property investment (see below).

A further problem is the difficulty in determining the indifference curves of an investor. In practice, it is possible to bypass the problem by defining a return or risk objective and determining the optimum portfolio to achieve that objective. However, the objectives of investors are more complex than optimisation of the risk/return trade-off from period to period. Modern fund management strategies are set against benchmarks and the MPT analysis has to be adjusted accordingly (see Section 7.4). Further, investors have to pay attention to their liabilities when determining the appropriate investment strategy (see Section 7.5).

Two further problems arise when applying MPT to property portfolios. First, the quality and availability of data from which to forecast returns and estimate standard deviations and correlations for assets are limited. In the UK and, to a lesser extent, in the USA, historical data are now available of sufficient quality, a long enough time series and appropriate disaggregations to overcome many of the problems. However, the difficulties of forecasting the inputs for individual properties mean that it is extremely unlikely that MPT could ever be used at that level of disaggregation in the property market (see Chapter 6).

Second, the procedure requires a period-by-period rebalancing of the portfolio to achieve the new optimum structure. This may lead to substantial buying and selling, the costs of which have to be included in the analysis. In the case of property, illiquidity may render the rebalancing impossible. More generally, whatever the asset class, substantial transactions each period are unlikely as fund strategies (as outlined above) are likely to be set with reference to a benchmark which will change little from period to period (see Section 7.4).

Property asset classes

In recent years, the issue of the appropriate asset classes in the property market has received much attention in the academic literature. Extensive reviews can be found in Eichholtz et al. (1995) and Hoesli and MacGregor (1995). The underlying purpose is to identify groupings within which property market returns are expected to be similar and among which returns are expected to be different.

In the USA, the main issue has been on the best geographical divisions to use in portfolio construction. Various classifications using economic data have been developed and tested (Corgel and Gay, 1987; Hartzell *et al.*, 1986; Shulman and Hopkins, 1988; Malizia and Simons, 1991; Mueller and Ziering, 1992; and Mueller, 1993). Similar work for Europe (Hartzell *et al.*, 1993) considered groupings based on employment structure. This work is limited by the absence of returns data, the reliance on data representing proxies for property demand and the omission of supply responses.

In the UK, researchers have been able to use property returns data which combine demand and supply factors (Hoesli and MacGregor, 1995; Hoesli, Lizieri and MacGregor, 1997; Hamelink *et al.*, 1999). The common theme is that the dominant pattern in groupings of local property markets is property type. None of the work was able to identify strong geographical or economic function dimensions to the local market returns. Jackson (1998) also found no clear geographical or functional dimension in her study of the UK retail property market. The conventional classification of the UK property market for portfolio construction is into three property types and 11 regions, plus various subdivisions of London depending on the property type. The appropriate classification remains an unresolved issue and has obvious links to the search for the appropriate level of analysis for property market forecasting (see Chapters 6 and 9).

Before turning to the practical application of MPT to property portfolios, it is first necessary to consider the Capital Asset Pricing Model (CAPM), which provides a theoretical method to calculate an appropriate risk premium for an investment. The next section also introduces the notion of specific risk which will be used in Section 7.4.

7.3 The Capital Asset Pricing Model

Modern Portfolio Theory (MPT) is an analytical framework for calculating the risk of a portfolio and for determining the optimal allocation to each asset. At the time of its development, the calculations and the data required made its use impracticable. The Capital Asset Pricing Model (CAPM) was developed in articles by Sharpe (1964), Lintner (1965) and Mossin (1966), in part, to overcome the problems of calculation involved in MPT, but also to develop and extend the analysis. The CAPM is a way of looking at the relationship between return and risk in a portfolio context. It considers the trade-off between return and risk to establish if an asset is correctly priced – in effect, it calculates an asset's correct risk premium.

The market model

The CAPM starts from the observation that most shares are positively correlated, that is, they tend to move upwards and downwards together. Sharpe suggested that this was the result of a common market response in their returns. This led to the proposition, known as the *market model*, that the expected return on a security could be expressed as a linear function of the expected return on the market as a whole:

$$E(R_i) = \alpha_i + \beta_i E(R_m)$$

◀Equation 7.9

where: $E(R_i)$ is the expected return on asset i
$E(R_m)$ is the expected market return
α_i and β_i are constants specific to the asset.

There is also a random error, giving:

$$R_{it} = \alpha_i + \beta_i R_{mt} + e_{it}$$
◄Equation 7.10

where: R_{it} is the return on asset i in period t
R_{mt} is the market return in period t
e_{it} is the value in period t of a random error, e_i, with zero mean and constant variance, σ_{ei}^2.

Using historical data it is possible to estimate the parameters by calculating the regression line:

$$R_i = a_i + b_i R_m$$
◄Equation 7.11

where: R_i is the historical return on the asset
R_m is the historical return on the market
a_i and b_i are estimates of α_i and β_i.

Thus, for any individual period, t:

$$R_{it} = a_i + b_i R_{mt} + r_{it}$$
◄Equation 7.12

where: R_{it} is the asset return in period t
R_{mt} is the market return in period t
r_{it} is the residual for asset i in period t.

The analysis can be extended to portfolios:

$$E(R_P) = \alpha_P + \beta_P E(R_m)$$
◄Equation 7.13

and

$$R_{Pt} = \alpha_P + \beta_P R_{mt} + \varepsilon_{Pt}$$
◄Equation 7.14

where: $E(R_P)$ is the expected portfolio return
R_P is the actual portfolio return in period t
$E(R_m)$ is the expected market return
R_m is the actual market return in period t
α_P and β_P are constants specific to that portfolio (these are weighted averages of the values of α_i and β_i of the individual assets which comprise the portfolio: see below)
ε_{Pt} is a random error for the portfolio in period t.

This approach assumes that the only common factor affecting returns is the return on the market and that there are no economic factors which affect only some sectors or some industries. This assumption can be relaxed in multi-factor models (see below). It has the advantage of simplifying the calculations by relating the

Figure 7.4► The market portfolio and the Capital Market Line

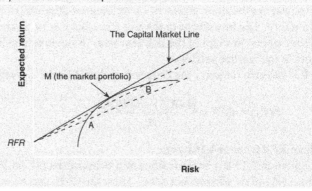

returns on any asset to a common index (the market as a whole) rather than to each other.

The Capital Market Line

To develop the CAPM model requires a number of assumptions (Alexander and Francis, 1986; Rutterford, 1993):

1. Investors are risk averse and maximise their expected rate of return over a single period.
2. Investors make their decisions solely on the expected return and standard deviation of portfolios.
3. The expected return and standard deviation of these portfolios exist.
4. All capital assets are infinitely divisible, so that parts of an asset can be bought. Investors are price takers (that is, no individual can affect the market by buying and selling). There are no taxes or transactions costs.
5. There exists a risk-free asset in which there can be unlimited investment or borrowing at the same rate. Inflation is fully anticipated in this rate.
6. All assets, including human capital, are marketable.
7. All information is free and simultaneously available to all investors.
8. All investors agree on the period under consideration and have identical expectations about risk.

An investor who conforms to conditions 1–4 would choose a portfolio on the efficient frontier under MPT. The other assumptions are required to build the CAPM model and their impact will be discussed later.

Figure 7.4 uses the efficient frontier constructed in Figure 7.1 and introduces a risk-free asset which yields a risk-free return (RFR). For investors A and B, their optimum portfolios can be combined with the risk-free asset to produce risk/return combinations as shown by the dotted lines. However, both investors could move to a higher indifference curve by combining the risk-free asset with the same portfolio, known as the *market portfolio*, on the efficient frontier. The line which joins the risk-free rate to the market portfolio is known as the *Capital Market Line* (CML). Thus, for any investor, the optimum portfolio is a combination of the common market portfolio and the risk-free asset. The particular combination preferred by an investor depends on their required risk and return. The investment in the risk-free asset can be either positive or negative. Money can be invested in the risk-free asset or funds can be borrowed to invest more in the market portfolio.

It is possible to derive the equation of the CML for a portfolio of assets:

$$E(R_\mathrm{P}) = RFR + \frac{[E(R_\mathrm{m}) - RFR]\sigma_\mathrm{P}}{\sigma_\mathrm{m}}$$

◄Equation 7.15

where RFR is the risk-free rate.

Equation 7.15 is a linear relationship between the risk (σ_P) and the expected return (E_P) of an efficient portfolio. Thus, the CML can be used to calculate the extra return which should be received as a reward for taking additional risk. The return is made up of the risk-free return (RFR) plus an extra return above the risk-free rate (($E(R_\mathrm{m}) - RFR)\sigma_\mathrm{P}/\sigma_\mathrm{m}$) which depends on the risk of the portfolio.[3]

Suppose the expected return on the market is 10%, the risk-free rate is 5%, the risk of the market is 20% and the risk of the portfolio is 10%, then:

$$E(R_\mathrm{P}) = RFR + \frac{[E(R_\mathrm{m}) - RFR]\sigma_\mathrm{P}}{\sigma_\mathrm{m}}$$

$$= 5 + \frac{(10 - 5)10}{20}$$

$$= 7.5$$

The expected portfolio return is 7.5 and the risk premium is 2.5.

The Securities Market Line

Equation 7.15 applies to portfolios lying along the CML. It is also possible to derive an expression for any investment. The equation is:

$$E(R_i) = RFR + \frac{[E(R_\mathrm{m}) - RFR]\mathrm{cov}_{i\mathrm{m}}}{\sigma_\mathrm{m}^2}$$

◄Equation 7.16

where $\mathrm{cov}_{i\mathrm{m}}$ is the covariance of asset returns with market returns.

The equation can also be written as:

$$E(R_i) = (1 - \beta_i)RFR + \beta_i E(R_\mathrm{m})$$

◄Equation 7.17

This is known as the *Securities Market Line* (SML) and it links the expected return on an investment to its risk. The SML is similar to the CML but rather than expected return being a function of risk as measured by the standard deviation, it is now a function of the covariance of the asset with the market. The term $\mathrm{cov}_{i\mathrm{m}}/\sigma_\mathrm{m}^2$ is known as the beta (β) of an asset. It measures the volatility of the asset *relative to the market*. The SML is shown in Figure 7.5. Note that the x-axis now measures β and not the standard deviation of returns.

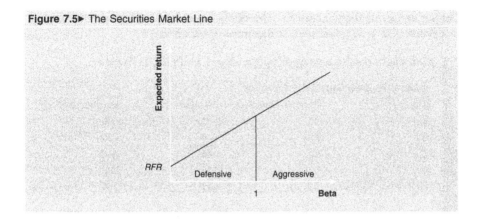

Figure 7.5► The Securities Market Line

▶ If beta is 1, then the asset has the same risk as the market and the expected return will be the same as the market expected return.

▶ If beta is zero, there is no risk and the expected return is the risk-free asset.

▶ If beta is less than 1, the expected return is less than the market expected return (this is known as a defensive asset).

▶ If beta is greater than 1, the expected return is greater than the market expected return (this is known as an aggressive asset).

From Equations 7.9 and 7.10, the risk of an asset (as measured by variance) is given by:

$$\sigma_i^2 = \beta_i^2 \sigma_m^2 + \sigma_{\varepsilon i}^2 \qquad\qquad \blacktriangleleft \text{Equation 7.18}$$

or (as measured by standard deviation, the positive square root of the variance):

$$\sigma_i = \sqrt{\beta_i^2 \sigma_m^2 + \sigma_{\varepsilon i}^2} \qquad\qquad \blacktriangleleft \text{Equation 7.19}$$

The risk of the asset, as measured by the variance (σ_i^2), may thus be divided into two parts. The first, $\beta_i^2 \sigma_m^2$, measures the risk relative to the market, and is known as *market* (or *systematic* or *non-specific* or *non-diversifiable*) risk. The second, $\sigma_{\varepsilon i}^2$, is specific to the asset and is known as *specific* (or *non-market* or *non-systematic* or *diversifiable*) risk.[4] For a portfolio with proportions, p_i, in n investments, the portfolio beta is the weighted average of the component betas:

$$\beta_P = \Sigma\, p_i \beta_i \qquad\qquad \blacktriangleleft \text{Equation 7.20}$$

and risk is:

$$\sigma_P = \sqrt{(\Sigma\, p_i \beta_i)^2 \sigma_m^2 + \Sigma\, p_i^2 \sigma_{\varepsilon i}^2} \qquad\qquad \blacktriangleleft \text{Equation 7.21}$$

Estimates of β and σ_ε based on historical data are available for major share markets. These are used as proxies for proper expectations values. For example, in the UK, the data are provided by the London Business School Risk Measurement

Service and are updated quarterly. This allows a portfolio's systematic and non-systematic risk to be calculated, as illustrated in Example 7.2.

Example 7.2

Consider a portfolio comprising five shares, as shown in Table 7.1.

Table 7.1▶ Systematic and specific risk

Share	Portfolio weight (p_i)	Beta (β_i)	Specific risk ($\sigma_{\varepsilon i}$)
A	0.2	1.10	0.40
B	0.3	1.25	0.30
C	0.2	1.20	0.20
D	0.2	1.05	0.40
E	0.1	0.95	0.25

The first stage is to calculate the portfolio's beta. From Equation 7.20:

$$\beta_P = \Sigma\, p_i \beta_i$$
$$= (0.2 \times 1.10) + (0.3 \times 1.25) + (0.2 \times 1.20) + (0.2 \times 1.05)$$
$$+ (0.1 \times 0.95)$$
$$= 1.14$$

Using a market risk of 20% as in Example 7.1, the portfolio systematic risk is given by:

$$\sigma_P = (\Sigma\, p_i \beta_i)\sigma_m$$
$$= 1.14 \times 0.2$$
$$= 0.228$$

The portfolio specific risk is given by:

$$\sigma_{\varepsilon P} = (\Sigma\, p_i^2 \sigma_{\varepsilon i}^2)^{1/2}$$
$$= \{[(0.2)^2 \times (0.4)^2] + [(0.3)^2 \times (0.3)^2] + [(0.2)^2 \times (0.2)^2] + [(0.2)^2$$
$$\times (0.4)^2] + [(0.1)^2 \times (0.25)^2]\}^{1/2}$$
$$= 0.152$$

The total risk is given by:

$$\sigma_P = [(\Sigma\, p_i \beta_i)^2 \sigma_m^2 + \Sigma\, p_i^2 \sigma_{\varepsilon i}^2)]^{1/2}$$
$$= [(0.228)^2 + (0.152)^2]^{1/2}$$
$$= 0.274$$

Note that the specific risk of the portfolio is smaller than that of any of the individual shares, but there remains a substantial amount of risk which could be diversified away by holding more shares. As the number of investments, n, increases and so each p_i decreases, the specific risk gets smaller. Thus, in a portfolio, the second component can be diversified away but the first cannot. This is a central aspect of the CAPM: an investor should not be rewarded for bearing risk which can be diversified but only for that which cannot. It will be used in Section 7.4.

The CAPM can be used to explain or attribute past performance.

Example 7.3

Consider the portfolio in Example 7.2. Suppose that last year the shares in the portfolio had the returns shown in Table 7.2.

Table 7.2▶ Delivered returns

Share	A	B	C	D	E
Return (%)	+15	−10	+20	+5	0

The portfolio return was the weighted average of the individual asset returns:

$$R_P = \Sigma \, p_i R_i \qquad \qquad \text{◀Equation 7.22}$$
$$= (0.2 \times 15) + (0.3 \times -10) + (0.2 \times 20) + (0.2 \times 5) + (0.1 \times 0)$$
$$= 5$$

This can be compared against a fully diversified portfolio with the same beta (1.14):

$$R_P = (1 - \beta_P)RFR + \beta_P R_m$$
$$= (1 - 1.14) \times 5 + (1.14 \times 10)$$
$$= 10.7$$

The difference between the portfolio return and that of a fully diversified portfolio with the same beta is the result of asset specific factors. In this example, these factors were detrimental and led to an *abnormal return* of −5.7%; in other years the abnormal return could be positive.

The CAPM as a pricing model

The CAPM, as the name suggests, is a pricing model. The market and the non-market components of risk have to be considered separately. In a market dominated by investors able to hold diversified portfolios, market risk assumes greater importance. As investors cannot diversify it away, but can diversify specific risk, market risk is the only risk which should be rewarded.

If the beta of the investment and the expected market return are known, the expected return on the asset can be calculated and compared against the return which the asset is priced to deliver. Figure 7.6 shows how the SML may be used to

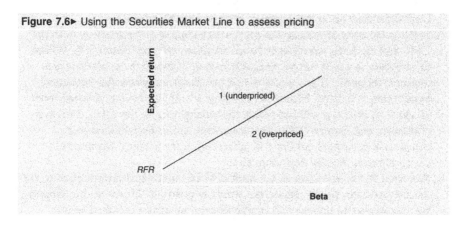

Figure 7.6▶ Using the Securities Market Line to assess pricing

assess the pricing of an investment. Asset 1 lies above the SML and so its expected return is above that required by its beta, so it is underpriced. In contrast, asset 2 lies below the SML and is priced to deliver a return below that required by its beta. Asset 2 is thus overpriced.

Consider an asset with a beta of 1.2. The risk-free rate is 5% and the expected return on the market is 10%. The correct expected return on the market is given by Equation 7.18:

$$E(R_i) = (1 - \beta_i)RFR + \beta_i E(R_m)$$
$$= (1 - 1.2) \times 5 + (1.2 \times 10)$$
$$= -1 + 12$$
$$= 11$$

Thus, the expected asset return is 11%. This can be compared against the results of analyses which estimate the return the asset is currently priced to deliver.

Practical limitations

The basic CAPM set out above is subject to a number of limitations:

1. *Single-period analysis*: To overcome the CAPM assumption of a single period, multi-period models have been developed but, as Radcliffe (1994: 259) suggests, 'they can always be criticized for certain assumptions that are required to make the model mathematically tractable'.
2. *Indivisibilities*: If assets are not infinitely divisible, investors cannot hold fractions of an asset and so small investors could not diversify. However, in practice, in the share market, particularly as it is dominated by large investors, this is not a problem. According to Alexander and Francis (1986), for shares held by many investors, the CAPM is an approximate equilibrium model; for shares held by a small number of investors, the share's standard deviation is a more relevant measure of their risk.
   This issue is of particular relevance in the property market where, generally, no two investors hold the same property and where lot size is large and so diversification is more difficult (see Chapter 2). Accordingly, the validity of CAPM for analysis of the property market must be limited.
3. *Transaction costs*: If transaction costs were included in the analysis, both the CML and the SML would have bands on either side (see Figure 7.7). Within these bands, it would not be profitable to buy or sell assets as costs would consume the profit. Thus, prices would not change to attain the theoretical equilibrium in CAPM. Moreover, investors would not have to diversify over all assets to obtain portfolios within the banded area on the CML. However, Alexander and Francis (1986: 136) argue that 'given their relative size, transaction costs need not have an effect that is particularly detrimental to the equilibrium picture derived in theory'.
4. *Tax rates*: If the same rate of tax applied to all capital and income returns, the market portfolio would remain the optimum portfolio. However, if a different tax rate applies to income and capital returns, an asset's expected return

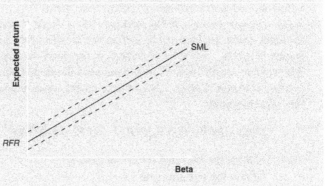

depends on both its beta and its income yield. If the income rate were higher than the capital rate, high yield assets would require higher pre-tax expected returns.

5. *The risk-free rate*: In the USA, where, until recently, there was no real risk-free investment, *zero-beta models* have been developed. In the absence of a risk-free asset, the lowest risk asset has zero beta.[5] In such a model, usually known as a *two-factor CAPM*, there are two sources of risk: the market return and the zero-beta portfolio return.

In practice, borrowing rates are higher than the investment rate. This means that the intercept in Figure 7.4 would vary depending on whether the investor wished to invest or borrow at the risk-free rate. The result is that the market portfolio (M) depends on whether an investor is investing or borrowing at the RFR.

6. *Non-marketable assets*: One of the assumptions of the CAPM is that all assets are marketable. However, many investors hold claim to non-marketable investments such as pensions and human capital which are relevant to investors when seeking to identify optimum portfolios. The former provide future cash flows while the latter produces wages. If these are included in the analysis, the conclusion is that investors should not all hold the same market portfolio. Moreover, another risk measure is required in addition to beta: the covariance with the portfolio of non-marketable assets. Another way of looking at the problem is to view the non-marketable assets as being marketable but having transaction costs which prohibit trading.

7. *Defining the market*: Linked to 6 above is a further limitation which is of relevance to the estimation of beta for property. Even if the conventional approach is taken of only considering marketable assets, the 'market' should include all investments, including property. In practice, however, it is usually taken to be the share market alone, thus making any estimate of property's beta unreliable.[6]

8. *Common expectations*: The basic CAPM assumes that all investors have the same expectations, so will all choose the same 'market' portfolio. If different groups were to have different views, each would identify a different optimum

portfolio. This would create additional sources of risk in returns as investors would choose their portfolios to reflect their views.

9. *Single common factor*: CAPM assumes only a single common factor in returns. However, there are likely to be factors which affect returns of particular groups of assets, such as sector effects in the share market.[7] To overcome this problem, multi-index models have been developed which link returns to other common factors. A two-factor model which includes inflation has the following form:

$$E(R_i) - RFR = \alpha_i + \beta_{i1}[E(R_m) - RFR] + \beta_{i2}E(IR)$$ ◄Equation 7.23

where: $E(R_i)$ is the expected return on asset i
RFR is the risk-free rate
$E(R_m)$ is the expected return on the market
$E(IR)$ is the expected value of the inflation rate
α_i, β_{i1} and β_{i2} are constants specific to investment i.

More general multi-factor models, called BARRA models after the firm which developed them, contain many more factors and are used by portfolio managers to attribute the sources of their returns. In the USA, some empirical studies have used multi-factor models with property data. Chan *et al.* (1990), for instance, use equity REIT data, while Ling and Naranjo (1998) use commercial property data.

A variant on multi-index models is Arbitrage Pricing Theory (APT) which, although apparently similar to multi-beta models, is based on very different economic theory.[8] These have the form:

$$E(R_i) = \alpha_i + \Sigma \, \beta_{ij}E(F_j)$$ ◄Equation 7.24

where: $E(R_i)$ is the expected return on asset i
$E(F_j)$ is the expected value of factor F_j
α_i and β_{ij} are constants specific to asset i.

While APT acknowledges the need to consider multiple risk factors, it does not identify these and different studies have used many different factors.

10. *Valuations*: The use of CAPM for property analysis is limited by the data. The need to use valuation data means that volatility is understated (see Chapters 2, 3 and 4) by valuation smoothing. Even then, there is likely to be limited time series data in most countries to calculate reliable estimates of beta. This problem becomes worse for individual properties where beta is unlikely to be temporally stable.

The problems outlined above mean that the practical application of the CAPM to the analysis of property investment is limited, and the procedure outlined in Chapter 5 is to be preferred as a pricing framework. Some limited attempts have been made, however, to use the CAPM with property data. These studies include Chan *et al.* (1990), Liu *et al.* (1990) and Brown (1991).

The next section considers the application of some of the theoretical ideas in MPT and CAPM to the practical management of property portfolios.

This section uses the basic MPT framework and the notion of specific risk from the CAPM to develop a practical means of constructing a property portfolio. Rather than absolute return and risk, it uses return and risk *relative to a market benchmark*.

In practice, fund objectives are not set in isolation but with reference to competitors or the market as a whole because, in a competitive market, fund managers would lose business if their performance were consistently worse than that of their competitors. If a fund management company is managing money on behalf of other organisations it must make competitive returns in order to retain or expand its business. Similarly, an insurance company must generate competitive returns for profit-related bonuses and to keep its fees competitive. One or two years' bad results may be enough for a manager to lose business. In such circumstances, objectives are set relative to a benchmark either of competitors or of the market. This approach to fund management has come to dominate property investment in the UK since the late 1980s when analyses of fund performance became available through the Investment Property Databank (IPD). The basic model is set out in this section and is applied in Chapter 9.

Adapting the basic MPT model, the expected return of a fund relative to a benchmark (in this case the market) is:

$$E(RR_P) = E(R_P) - E(R_m)$$
$$= \Sigma\, w_i E(R_i) - \Sigma\, W_i E(R_i)$$
$$= \Sigma\, (w_i - W_i) E(R_i) \qquad \qquad \text{◄Equation 7.25}$$

where: $E(RR_P)$ is the expected relative return for the portfolio
$E(R_P)$ is the expected portfolio return
$E(R_m)$ is the expected market return
$E(R_i)$ is the expected return for asset i
w_i is the proportion of the portfolio in asset i
W_i is the proportion of the market in asset i.

Note that:

$$\Sigma\, w_i = \Sigma\, W_i = 1 \qquad \qquad \text{◄Equation 7.26}$$

so:

$$\Sigma\, (w_i - W_i) = \Sigma\, w_i - \Sigma\, W_i = 0 \qquad \qquad \text{◄Equation 7.27}$$

The expected portfolio relative risk is known as the *tracking error* and is defined as:

$$TE_P = \sqrt{\Sigma\Sigma (w_i - W_i)(w_j - W_j)\sigma_i\,\sigma_j\,\rho_{ij}} \qquad \qquad \text{◄Equation 7.28}$$

Equations 7.25 and 7.28 are equivalent to Equations 7.1 and 7.3 with w_i replaced by $(w_i - W_i)$. Note that if the portfolio has the same structure as the market – that is, it tracks the market perfectly – so $w_i = W_i$ for all i, the relative return and relative risk are both zero. Thus, even if the benchmark has high *absolute* risk and the portfolio has the same high risk structure, its *relative* risk is zero.

This basic approach is used in other investment markets but has to be modified for the particularities of property. An additional term is required for the specific risks of individual properties. In constructing a portfolio using MPT, it is generally assumed that each asset class contains only systematic risk and that the specific risk can be diversified away. While such diversification is relatively easy in the share market, in the property market, large lot size and heterogeneity make diversification difficult except for very large portfolios.

Suppose the return on an individual property comprises two components: a return attributable to the asset class (i) to which it belongs; and a return specific to the individual property (k).

$$R_k = R_i + R_{STk}$$

where: R_k is the asset return
$\quad\quad R_i$ is the asset class return
$\quad\quad R_{STk}$ is the property specific (stock) return for property k.

For the market as a whole, R_{STk} has an expected value of zero and a standard deviation of σ_{STk}. Each R_{STk} is independent of any other R_{STk} and of R_i, thus:

$$E(R_k) = E(R_i) \quad\quad\quad\quad\quad \blacktriangleleft\textbf{Equation 7.29}$$

and:

$$\sigma_k^2 = \sigma_i^2 + \sigma_{STk}^2 \qu\quad\quad\quad\quad \blacktriangleleft\textbf{Equation 7.30}^9$$

where: σ_k is the total property risk
$\quad\quad \sigma_i$ is the asset class risk
$\quad\quad \sigma_{STk}$ is the property specific risk.

When the individual properties are combined within a portfolio, the asset class components of risk (σ_i^2) combine as in Equation 7.3. The property-specific variances are additive as they are independent. These latter risks add a new term to the portfolio risk and portfolio relative risk equations:

$$\sigma_P = \sqrt{\Sigma\Sigma w_i w_j \sigma_i \sigma_j \rho_{ij} + \Sigma p_k^2 \sigma_k^2} \quad\quad \blacktriangleleft\textbf{Equation 7.31}$$

$$TE_P = \sqrt{\Sigma\Sigma (w_i - W_i)(w_j - W_j)\sigma_i \sigma_j \rho_{ij} + \Sigma p_k^2 \sigma_k^2} \quad\quad \blacktriangleleft\textbf{Equation 7.32}$$

where p_k is the proportion of the portfolio in property k.

The new term depends on the number of buildings and their relative sizes. For illustration, suppose that the property-specific risks are equal, so $\sigma_k = \sigma$ for

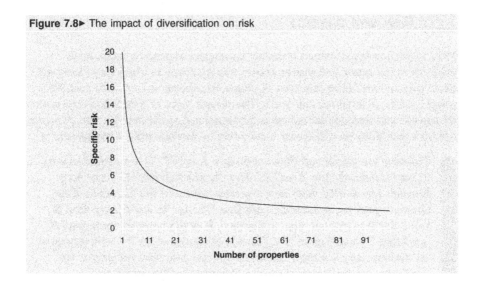

all k; and that the proportions in each property are equal, so that $p_k = 1/n$ for all k, then:

$$\Sigma\, p_k^2 \sigma_k^2 = \Sigma \left(\frac{1}{n}\right)^2 \sigma^2$$

$$= n \left(\frac{1}{n}\right)^2 \sigma^2$$

$$= \frac{\sigma^2}{n} \qquad\qquad \text{◄Equation 7.33}$$

Thus, as the number of properties increases, the portfolio-specific risk decreases. Assuming a value of 20 for σ, this is illustrated in Figure 7.8. Up to about 15 properties the risk reduction is substantial, thereafter it is much less.

In the share market, if the asset classes were defined as individual shares, there would be no need for the second term in Equation 7.32. Further, for any reasonably sized portfolio, it would be possible to hold virtually the same weights as the benchmark in identical investments, so the tracking error would be effectively zero. If asset classes (such as industrial sectors) were used, and benchmark weights were held in these, the second term could be reduced to a negligible amount for even very small portfolios by holding a reasonable number of shares within each asset class. However, it is not possible to hold the same properties as a market benchmark and the second term is needed to take account of the unique properties in a portfolio. Moreover, large lot size makes diversification more difficult. Thus, even with benchmark weights, the portfolio may retain significant specific risk. These issues are considered again in the context of a notional portfolio in Chapter 9. The next section considers another dimension to risk: liability risk.

Most major investors, such as insurance companies or pension funds, have liabilities in the future and seek to ensure that the assets in which they invest will allow them to meet these liabilities. A simple investment strategy is to match the present values of liabilities and assets. The present value of a cash flow depends on the amount and timing of the individual payments and on the discount rate. However, different cash flows have different sensitivities to changes in the discount rate.

Consider two simple cash flows: cash flow A is £417.74 to be received after 15 years; and cash flow B is £1083.50 to be received after 25 years. At a discount rate of 10%, both have a present value of £100. However, if the discount rate were to increase, cash flow A would be worth more than B, and if the discount rate were to decrease, B would be worth more than A (see Table 7.3 and Figure 7.9). The rate of change of the PV with respect to the discount rate is termed the *duration* of the cash flow: the greater the sensitivity and the steeper the curve, the greater is the duration.

Table 7.3▶ The sensitivity of cash flows to changes in the discount rate

Discount rate	PV of A	PV of B
5.0%	200.9	320.0
6.0%	174.3	252.5
7.0%	151.4	199.6
8.0%	131.7	158.2
9.0%	114.7	125.7
10.0%	100.0	100.0
11.0%	87.3	79.8
12.0%	76.3	63.7
13.0%	66.8	51.0
14.0%	58.5	40.9
15.0%	51.3	32.9

Figure 7.9▶ The sensitivity of cash flows top changes in the discount rate

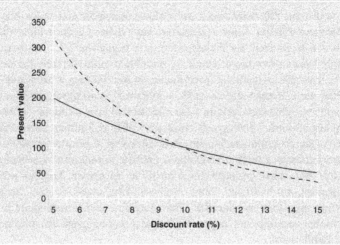

Figure 7.10► Duration of liabilities greater than that of assets

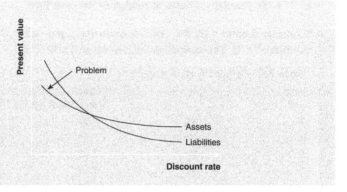

Figure 7.11► Duration of liabilities less than that of assets

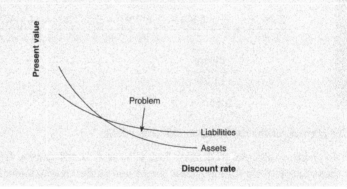

Suppose a fund matched the present values of its assets and liabilities. If the duration of a fund's liabilities were greater than that of its assets (see Figure 7.10), and the discount rate were to fall, the assets would have a lower present value than the liabilities. This would create problems. Similarly, if the liabilities had lesser duration than the assets, and the discount rate rose, there would be problems (see Figure 7.11). However, if the durations of the assets and liabilities were the same, the present values of assets and liabilities would change in the same way and the fund would be immunised whether discount rates fell or rose. Duration adds a new dimension of risk. Like tracking error, it can be viewed as a form of relative risk, in this case relative to liabilities.

Most work has been undertaken on the duration of bonds as the cash flows are known in advance. For bonds it is possible to calculate duration as the average time to the receipt of each cash flow, weighted by the present value of each cash flow, thus:

$$\frac{\Sigma PV_t Y_t}{\Sigma PV_t}$$

◄**Equation 7.34**

where: PV_t is the present value of the cash flow in year t

Y_t is the number of years to receipt of the cash flow.

Example 7.6

Consider a bond with five years to maturity, a par value of £100 and a coupon of £10. The cash flows are shown in Table 7.4.

Table 7.4▶ Calculating bond duration

Year	Cash flow	Present value of cash flow at 12.5%	Year × PV
1	10	8.89	8.89
2	10	7.90	15.80
3	10	7.02	21.07
4	10	6.24	24.97
5	110	61.04	305.21
Total		91.10	375.94

The duration of this bond is:

$$\frac{\Sigma PV_t Y_t}{\Sigma PV_t} = \frac{(1 \times 8.89) + (2 \times 7.90) + (3 \times 7.02) + (4 \times 6.24) + (5 \times 61.04)}{8.89 + 7.90 + 7.02 + 6.24 + 61.04}$$

$$= \frac{375.94}{91.10}$$

$$= 4.13$$

The general results for bonds are (Ward, 1988):

▶ As coupon falls, duration rises. This is because as the coupon falls the importance of the final payment grows and so the weighted average time to payment increases.
▶ As term rises, duration rises. This is because it takes longer to get the payments.

The conventional view on the durations of the main UK asset classes is shown in Table 7.5. In this, property has a duration between that of bonds and shares, reflecting its hybrid bond and equities characteristics. However, some researchers have identified a problem. The simple formulae for duration are derived from the bond market where cash flows are fixed and so do not change with the discount rate. However, for shares and property, if a change in the nominal discount rate were the result of a change in inflation expectations, the expected cash flows would

Table 7.5▶ Durations of main UK asset classes

Asset class	Duration (years)
Conventional bonds	7
Index linked bonds	15
Property	17
Shares	20

Source: Ward (1999).

also change. The greater is the resultant change, the greater is the reduction in the duration from the estimates shown in Table 7.5: Hamelink *et al.* (1998) suggest values of below 10 for shares and property.

Duration has been used primarily by actuaries who have to ensure that institutions can match their liabilities. Although it is of no practical relevance to the management of property portfolios, it helps explain the broad allocations to the main asset classes as discussed in Chapter 10 and the differences in allocation among different types of investors as shown in Chapter 2.

7.6 Summary and conclusions

This chapter has dealt with risk in the context of investment portfolios. It has reviewed conventional finance theory with a view to applying this to the construction of a property portfolio. Two main areas of theory were considered: Modern Portfolio Theory (MPT) and the Capital Asset Pricing Model (CAPM). Tracking error and duration were also discussed. The former has substantial practical importance but the latter has no direct relevance to property portfolios.

When assets are combined in a portfolio, although the expected portfolio return is simply the weighted average of the individual asset expected returns, portfolio risk is not a simple weighted average. Rather it depends on the correlations among the assets: other things being equal, the lower are these correlations, the lower is the portfolio risk. MPT provides a theoretical model for efficient diversification. It involves the following stages:

► Estimate the individual asset expected returns and risks.
► Calculate the possible combinations of portfolio expected return and risk from all possible combinations (weights) of the assets.
► Eliminate the inefficient portfolios to establish the efficient frontier.
► Determine the optimum portfolio for an individual investor by consideration of their return/risk indifference.

The primary problem with MPT when applied to individual investments is the need to estimate the expected return and risk for each investment and the correlations between investments. Historical data may be used but are of little value for estimating expected returns. To an extent this can be overcome by considering assets classes, that is, groups of investments with similar characteristics. Valuation-based property data create additional problems because of the smoothing and understatement of risk.

Although it is extremely difficult to determine return/risk indifference, this can be overcome in practice by requiring a clear return objective and identifying the most efficient portfolio to achieve this. The likely rebalancing of a portfolio from period to period creates further difficulties for an illiquid asset such as property.

The CAPM was developed, in part, to overcome some of the data requirements of MPT. In the CAPM, a risk-free asset is introduced and borrowing or investment allowed at the risk-free rate. The conclusion of the CAPM is that all investors should invest in the same 'market' portfolio of risk assets plus an investment in (or loan of) the risk-free asset. The expected return and risk for the investor can be adjusted by changing the proportions invested in the market portfolio and the risk-free asset.

The CAPM also provides a pricing model in which risk is divided into two parts: a market component which cannot be diversified away; and a component specific to an investment which can be diversified. Investors should be rewarded only for the former as the latter can be avoided by diversification. The CAPM also introduces 'beta', a measure of the correlation of an investment with the market, which is used to estimate the appropriate risk premium of an investment. The model can be used to attribute historical performance to market effects and to investment-specific effects, and to consider mispricing in the market.

However appealing it might seem as a model, the CAPM suffers from a number of limitations, including:

- ▶ It is a single-period model.
- ▶ It assumes that investments are infinity divisible – a particular problem for property.
- ▶ It omits costs and taxes.
- ▶ Risk-free borrowing and lending rates are not the same.
- ▶ Non-marketable assets should be included in the analysis.
- ▶ It assumes that all investors have the same expectations.
- ▶ It assumes that there is a single common factor in returns.
- ▶ Valuations are likely to understate the beta of property.

Several of these have been overcome by developments of the CAPM; for others, the basic conclusions derived from the assumptions have been shown to be reasonably robust. However, there are great practical limitations in applying the CAPM to the property market, particularly at the level of the individual property. The method outlined in Chapter 5 is likely to be a more fruitful way of considering mispricing.

Despite their limitations, MPT and the CAPM can be adapted to practical portfolio management. Fund managers are generally concerned not with their absolute performance but with their performance relative to their competitors. A variant of MPT was developed which uses return and risk relative to a competitive benchmark and which incorporates analysis of specific risk. In this, the new measure of risk is tracking error: a property portfolio with a high absolute risk can have almost no tracking error risk if it has the same structure as the market and is well diversified. This will be applied in Chapter 9 to practical property portfolio management.

Finally, the chapter considered liability risk, that is, the risk that assets may not be able to meet the investor's liabilities. Duration was defined as the sensitivity of the present value of an asset (or liability) to a change in the discount rate. When durations of assets and liabilities are matched, an investor is immunised against changes in the discount rate. Duration is a consideration at the mixed-asset level and has limited relevance to property fund management. However, it offers insights into the allocations to assets within a mixed-asset portfolio.

The final component of the pricing model introduced in Chapter 5 is depreciation, which is considered in Chapter 8. Chapter 9 then considers practical property portfolio management and Chapter 10 considers the role of property in the mixed-asset portfolio.

Notes

1. The exception is an 'artificial' paper asset which may be constructed to be perfectly positively or negatively correlated with a real asset.
2. It is assumed that the joint probability distribution of R_{it} and R_{mt} is stationary and bivariate normal. Thus, the error term has a mean of zero and a constant variance and is uncorrelated with the return on the market. It is also assumed that there is no serial correlation in the error terms.
3. Three standard CAPM performance measures can be calculated for portfolios. The Sharpe measure is the difference between the portfolio return and the risk-free rate, divided by the portfolio risk. The Treynor performance measure is the difference between the portfolio return and the risk-free rate, divided by the portfolio beta. The Jensen alpha is a measure of the return a managed portfolio earned above that of an unmanaged (buy and hold) portfolio with equal market (systematic) risk. As these are not used in this book, they are not considered further. For a full discussion see Elton and Gruber (1995), Chapter 24.
4. Note that, while the variances (the standard deviations squared) are additive, the standard deviations are not.
5. In the UK, index-linked gilts provide a real risk-free rate.
6. See also Liu *et al.* (1990).
7. This is linked to the issue of non-marketable assets where two factors affect the risk of an asset: the covariance with the market and the covariance with the non-marketable assets.
8. Although APT appears very similar to the CAPM, the theories have very different bases. CAPM is based on utility theory, while APT is based on the principle of arbitrage. APT does not assume that investors are interested only in expected returns and standard deviations, nor does it have anything to say about the optimality of the market portfolio (see, for example, Radcliffe, 1994, for a fuller discussion).
9. This is equivalent to Equation 7.18 with $\beta = 1$ as the individual asset returns vary in the same way as the market of which they are part.

Further reading

Bodie, Z., Kane, A. and Marcus, A. J. (1993) *Investments*, Irwin, Burr Ridge (IL) (second edition): Chapters 5, 6, 7, 8, 9, 10, 11 and 15.

Brown, G. R. (1991) *Property investment and the capital markets*, E & FN Spon, London: Chapters 1, 4 and 5.

Elton, E. J. and Gruber, M. J. (1995) *Modern portfolio theory and investment analysis*, John Wiley & Sons, Inc., New York (fifth edition): Chapters 4, 5, 6, 7, 8, 10, 13, 14, 15, 16 and 21.

Fabozzi, F. J. and Modigliani, F. (1992) *Capital markets: institutions and instruments*, Prentice Hall International Editions, Englewood Cliffs (NJ): Chapters 4, 5 and 12.

Farrell, J. L. (1997) *Portfolio management: theory and application*, McGraw-Hill International Editions, New York (second edition): Chapters 2, 3, 4 and 5.

Ling, D. C. and Naranjo, A. (1998) The fundamental determinants of commercial real estate returns, *Real Estate Finance*, 14(4), 13–24.

Markowitz, H. M. (1959) *Portfolio selection: efficient diversification of investments*, John Wiley & Sons, New York.

Sanders, A. B., Pagliari, J. L. and Webb, J. R. (1995) Portfolio management concepts and their application to real estate, in *The handbook of real estate portfolio management*, Pagliari, J. L. (Ed.), Irwin, Burr Ridge (IL), 117–72: Chapter 2.

Sharpe, W. F., Alexander, G. J. and Bailey, J. V. (1995) *Investments*, Prentice Hall International Editions, Englewood Cliffs (NJ) (fifth edition): Chapters 7, 8, 9, 10, 11, 12 and 16.

Chapter 8

Depreciation

Introduction

Depreciation was introduced in Chapter 5 as loss of rental income (and hence capital value) of an ageing property when compared to an equivalent new property. It was shown that, for a correctly priced property, the capitalisation rate could be approximated as the sum of the risk-free rate, the risk premium and the depreciation rate, less the expected income growth. Thus:

$$k = RF + RP - G + d$$ ◄Equation 8.1

where: k is the capitalisation rate
   RF is the risk-free rate
   RP is the risk premium
   G is the expected rental growth
   d is the expected depreciation.

Explicit discounted cash flow (DCF) pricing analysis requires forecasts of each of the components of the capitalisation rate. Chapter 6 considered how to forecast G, then Chapter 7 looked at RP. This chapter now considers d, depreciation. As shown in previous chapters, significant progress has been made in rental forecasting and in risk analysis. In contrast, depreciation has received more limited attention, although this has been growing. There are a number of possible reasons for both the original lack of research and the subsequent growth in interest:

1. Buoyant markets and high inflation, which result in high levels of nominal rental growth, have concealed the effect of depreciation. Thus, it tends to become an issue when its impact is more obvious, in poor markets when rents fall and vacancy rates rise, and in periods of low inflation.
2. The persistence of implicit valuation methods means that depreciation is not considered explicitly. It should, however, be remembered that any valuation includes an estimation of depreciation, albeit implicitly. The growing use of explicit techniques has led to renewed interest in the topic.

3. Depreciation is not directly important to paper assets such as shares and bonds (although it is an important feature in tax accounting), but is important to direct property holdings. Accordingly, unlike income forecasting and risk analysis, it has not benefited from research by other investment professionals and academics.[1]
4. In the UK, many owners thought (perhaps hoped) that the standard full repairing and insuring (FRI) lease protected them against depreciation as it was largely the tenants' responsibility. However, the tenants' obligations to repair cover only a part of what is collectively known as depreciation. As technological change accelerates, so do occupiers' requirements, and the life-span of many commercial properties decreases through obsolescence. As this happens, the impact of depreciation on the value, and so on the owner, becomes more pronounced.

Of all the inputs into the valuation framework, depreciation has the least precise meaning in common usage. For example, the terms 'depreciation' and 'obsolescence' are often used interchangeably. Thus, the discussion begins, in Section 8.2, with a consideration of definitions of depreciation based on its effects. A proper understanding of the phenomenon requires examination of its different causes. These are considered in Section 8.3 and are used to classify depreciation. From a practical perspective, it is important to be able to measure the level of depreciation. This is examined in Section 8.4. Finally, Section 8.5 provides the summary and conclusions.

8.2 Defining depreciation by its effects

There are two broad ways to define depreciation according to its effects: one with reference to value, the other with reference to stock loss.

Depreciation as a fall in rental or capital value

In this definition, depreciation is 'a loss of rental or capital income of an ageing property when compared to an equivalent new property'. Thus, the rate of rental depreciation during period t is given as:

$$d_{it} = 1 - \frac{R_{it}/R_{it-1}}{R_{nt}/R_{nt-1}}$$

◄Equation 8.2

where: d_{it} is the rate of depreciation for property i (the subject property) in period t
R_{it} is the rental value of the subject property at the end of period t
R_{it-1} is the rental value of the subject property at the start of period t
R_{nt} is the rental value of an equivalent new property (the reference property) at the end of period t and
R_{nt-1} is the rental value of an equivalent new property at the end of period $t - 1$.

Table 8.1 shows rental indices for the subject and a continually new property. Depreciation is calculated using Equation 8.2.

Table 8.1▶ Calculating depreciation as a loss of rental value

Year	Rent of new property	Rent of subject property	Depreciation
0	100	100	
1	105	103	1.9%
2	111	107	1.7%
3	117	110	2.5%
4	123	114	1.4%
5	130	118	2.1%

As shown in Equation 8.1, depreciation also affects the capitalisation rate. The higher the rate of expected depreciation (for the life of the property), the lower would be the income stream and, thus, the lower the capital value and the higher the capitalisation rate.

Depreciation can also affect the capitalisation rate in another way. The expected depreciation is an estimate with a range of possible values. The greater this range – that is, the greater is the risk associated with the estimate – the higher should be the risk premium and the higher the capitalisation rate. It is possible that the depreciation estimates for older properties, for example of a particular type of construction, may be riskier than for newer properties. As a property ages, its specification becomes further from that of a modern property, so the risk of voids increases, and the risk premium and the capitalisation rate should rise. Thus, although it is possible to consider depreciation as a loss of capital value, this includes both the *income* and the *capitalisation rate* effects.

Note that this formulation assumes a compound or geometric *rate* of depreciation. If this rate were constant over time, it would produce a pattern of rental value *relative* to a new property, as shown in Figure 8.1. By contrast, in accountancy, it is usual to assume that depreciation results in the loss of a constant *amount* of the *initial value* of an asset. Thus, its value follows the linear pattern shown in Figure 8.2. This approach is applied to assets with relatively short lives, thus an asset with an *assumed* five-year life will decline by 20% of its

Figure 8.1▶ Compound depreciation measured relative to a new property

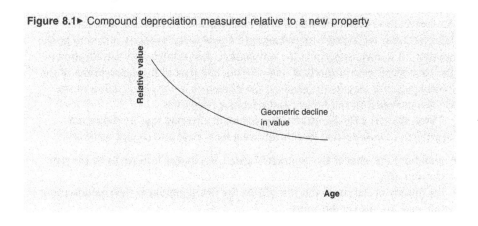

Geometric decline in value

Relative value

Age

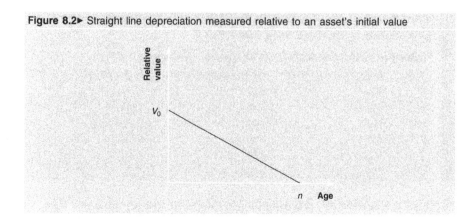

Figure 8.2▶ Straight line depreciation measured relative to an asset's initial value

initial value each year for five years. At the end of the five years, its 'book' value would be zero. This has rather more to do with accountancy conventions than with actual market values, and so is not a helpful approach for property analysis.

Defining an equivalent new property

The formula for depreciation given in Equation 8.2 requires a definition of an 'equivalent' new property. This creates a number of practical problems.

First, it is clear that the equivalent property must be in the same *sub-market* and of the same *type*. Thus, a new office in Aberdeen would not be 'equivalent' when considering a two-year-old property in the City of London, even if they were of identical construction. Similarly, a low-density campus-style office development would not be equivalent to a high-rise office block. While these are extremes, the appropriate sub-market, both spatially and by property type, cannot be defined precisely. This is a standard problem when considering property sub-markets for any purpose. It is exacerbated by changes to sub-market boundaries as a result of changes in relative prices.

Second, the *specification* of an equivalent new property creates difficulties. The quantity and quality of the services provided by a new property are both likely to increase over time as building technology and user demands change. Thus, an equivalently specified new property is unlikely to exist and, if it did, the depreciation measured would omit obsolescence. In practice, therefore, depreciation is measured with reference to a new *prime* property, taken to be the top rent for a new property in the sub-market. Accordingly, the specification of the 'equivalent' new property is time-varying and part of the depreciation of the subject property may be the result of the difference in the composition of the services provided by the subject and reference properties.

Third, there is a further practical problem in ensuring that the leases are 'equivalent'. Lease-related factors affecting rent, yield and capital value are:

▶ quantum: the smaller the contracted space, the higher is likely to be the rent per unit area
▶ the quality of the covenant: this affects the risk premium in the capitalisation rate and, so, the capital value

- the likely differences between rents agreed at review (and so based on market evidence) and those based on open market letting (which create new market evidence)
- incentives in weak markets also need to be taken into account so that *effective* rent is considered
- non-market leases between, for example, a parent and a subsidiary company.

In practice, older properties with a short lease and a high-risk tenant are unlikely to have an 'equivalent' new property. Such factors could be considered in estimates of depreciation but typically are not.

Depreciation as a fall in stock

The second definition of depreciation does not refer directly to value. It is 'the proportion of a stock which is removed from the market'. Thus:

$$\delta_t = \frac{TS_{t-1} - SS_t}{TS_{t-1}}$$

◄Equation 8.3

where: δ_t is the rate of depreciation of the stock during period t
TS_{t-1} is the total stock at the start of the period (including new developments during period $t - 1$)
SS_t is the stock from the start of the period which survives to the end of the period (that is, it excludes additions to the stock during period t).

This is the definition of depreciation used in models of office markets (see Chapter 6 and Ball *et al.*, 1998). Example 8.2 and Table 8.2 illustrate the calculation.

Table 8.2► Calculating depreciation as a loss of stock

Period	Total stock (m²)	New development	Surviving stock	Loss of old stock	Depreciation
0	2 500 000				
1	2 510 000	50 000	2 460 000	40 000	1.6%
2	2 540 000	75 000	2 465 000	45 000	1.8%
3	2 530 000	25 000	2 505 000	35 000	1.4%
4	2 500 000	5 000	2 495 000	35 000	1.4%
5	2 455 000	5 000	2 450 000	50 000	2.0%

The two definitions of depreciation are linked, for, when a property's capital value falls to a certain level, it will be removed from the market and its site may be redeveloped.

Fluctuations in the rate of depreciation

The latter definition introduces the important issue of the 'cohort' or 'vintage' of the stock. New stock is built at a variable rate, depending on demand, and is of different types and qualities of construction as user demands and technology change. Demand cycles are an explanation of the first factor; and the 1960s style of office construction is an example of the second. Thus, the rate of depreciation of stock is unlikely to be constant and may be prone to significant fluctuation as whole vintages of stock become obsolete together. This creates fluctuations in the requirement for new stock and so, in turn, in future depreciation.

Substantial falls in demand may increase the rate of depreciation as falls in rent would mean a filtering upwards in the occupation of space. On the other hand, a rise in rents would lead to a filtering down to lower quality space. As above, this creates problems of defining the sub-market within which to define depreciation.

The above discussion of definitions of depreciation has, in passing, raised a number of issues relating to different causes of depreciation. These are considered in the next section.

8.3 Classifying depreciation by its causes

The previous section considered two ways of defining depreciation by its affect. This section extends the discussion to a consideration of the underlying causes of depreciation.

The loss in the rental or capital value of a property, or its removal from the stock, is the result of a loss of utility or productive capacity. There are two broad categories of causes: *deterioration* which is directly linked to the passage of time and likely to have a relatively constant impact; and *obsolescence* which is not directly associated with the passage of time and is likely to have a variable time pattern of impact. The discussion considers, first, the building (that is, the physical structure) and then the site (that is, the land).

Building deterioration

There are two broad causes of deterioration, both of which are linked to the passage of time: wear and tear through *use*; and what has been variously termed 'the weather', the 'actions of the elements' or *environmental factors*. A property loses utility as it wears out through use; the rate at which this occurs is linked to the intensity of use. Environmental factors also result in a loss of utility. These include: water from rain, snow, ice and condensation; sunlight; temperature; wind; and air contamination. They cause construction materials or building components to lose functionality.

Both forms of deterioration are unavoidable although the quality of design and construction may lessen the impact. To a significant extent, both may be overcome by expenditure on either regular maintenance or periodic refurbishment.[2] This expenditure, as much as any loss in rent or capital value, is a cost of depreciation. Information on such expenditure can be used to forecast the likely future impact. Using the terminology of CAPM in Chapter 7, this form of depreciation may be regarded as *systematic* depreciation as, to some extent, it affects all properties.

Building obsolescence

Obsolescence may be generally defined as a decline in utility not directly linked to physical usage or to environmental factors. It is, therefore, not directly linked to the passage of time. However, the age or vintage of a building provides information about its type and quality of construction, both of which are important factors in obsolescence. This form of depreciation is likely to vary substantially over time, and particular forms of construction are likely to be affected significantly during relatively short periods.

There are four sub-categories: obsolescence of the plant (such as the functional capacity or efficiency of air conditioning, heating, electrical services, water and lifts); functional obsolescence of the structure (such as floor to ceiling height and flexibility of the floor plan); aesthetic obsolescence of the structure as a result of changes in taste; and, at least in the case of multi-let properties, absence of complementary services, such as dining and conference facilities. Each involves a mismatch between the services that a property can provide and the requirements of occupiers. This mismatch is a result of changes in both demand and supply. Changing user requirements, whether as a result of changing business practices or tastes, result in changes to the services required; while changing building technology results in a change to the services that can be provided. The mismatch between demand and supply of services results in a decline in property utility and so to a decline in value.

Although, there is no clear boundary, most forms of plant and aesthetic obsolescence and the absence of facilities can be overcome by expenditure, but most forms of functional obsolescence cannot. The boundary becomes less distinct when the cost of remedial action is considered: the expenditure may not be justified by the consequent increase in value.

The occurrence of some aspects of obsolescence, particularly to plant, is predictable, in the sense that technological progress can be anticipated, but the form of such progress is not and, consequently, the timing and impact of obsolescence are not predictable. Other aspects are much more difficult to anticipate. Changing office organisation led to the obsolescence of many 1960s properties as they had structural pillars which inhibited easy adaptation for open plan offices. Many of these properties also suffered from their restricted floor to ceiling heights which meant that electrical cabling, associated with expanded computer usage, could not be accommodated. However, recent developments in much less bulky fibre optic technology have meant that such properties are now able to accommodate the cabling.

This form of depreciation is unlikely to affect all properties at one time. Rather it is likely to have an affect on a particular type or construction of building.

Site depreciation

Most of the depreciation literature focuses on the physical structure and, so far, discussion here has been restricted to this.[3] However, much of the utility of a property comes from the location of its site. Although in practice, the structure is inseparable from its site, it is, nonetheless, useful for a fuller understanding of depreciation to consider the separation conceptually, as the site, too, can be affected by depreciation.

Prices in the land market reflect the demand for and supply of land for different uses. As a substantial part of the value of a property is its site, property values are affected by the land market. An increase in the value of the site, because of the demand for an alternative use, could also lead to its redevelopment. It is useful to divide changes in land prices into four broad categories.

1. *Changes affecting the entire market*: These are the result of systematic or market factors, such as a general economic downturn. Conceptually, such factors are best considered under income change (G) rather than depreciation (d). However, there are likely to be differential effects, in location, type of property or vintage of building, as occupiers 'filter' upwards or downwards between sub-markets as a result of differential price changes. As demand falls, some properties with poorer location or specification become redundant. The *relative* effects are, therefore, important when measuring depreciation.
2. *Changes affecting a particular type of use*: The changing locational preferences of particular users affect the utility of particular types of location. For example, the decline and movement of manufacturing industry from the 1960s onwards left large areas of vacant land in many European and US cities. The consequent change in the local environment had a re-enforcing effect on the decline in land values. More recently, the expansion of out-of-town shopping has resulted in demand for large peripheral sites and some cities have seen a consequent decline in their central areas. These relative price changes can be regarded as a form of obsolescence.
3. *Changes affecting a particular area or location*: These may be linked to 2 above and are the result of movement from or to an area. For example, as an area develops as an office location, complementary facilities such as shops and restaurants are likely to be set up. This makes the area more attractive to office users. A variant of this is the concentration of particular types of office users in an area. Here, location close to similar users makes the area attractive. Alternatively, an area in decline may benefit from some form of government intervention, such as infrastructure investment (particularly on roads) or incentives to locate in particular areas. The imposition or lifting of planning restrictions can have an effect, as can contamination and pollution. Again, the relative price changes can be regarded as a form of obsolescence.
4. *Changes affecting only a particular site*: It is rather more difficult to identify factors which affect only one site. Possible examples are planning restrictions on redevelopment of the site and the discovery of localised contamination or ground conditions, such as old mineral workings. These relate to previous site uses and are a form of site deterioration.

Categories 2 and 3 above are examples of site obsolescence, although some aspects may result in an *increase* in relative utility. These are systematic in that they affect all land in a particular use or location. Category 4 is a form of site deterioration – that is, the impact of wear and tear – and is much less common. However, contamination from previous uses, typically specific to one large site or several adjoining sites, has become of greater importance during the last 20 years.

The classification presented in this sub-section is summarised in Figure 8.3.

Other distinctions

A number of important distinctions arise from the above discussion.

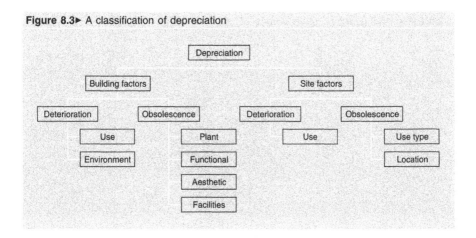

Figure 8.3▶ A classification of depreciation

1. *Predictability*: Building deterioration can be predicted and sensible estimates made of its impact by, for example, considering the costs of maintenance. The allocation of the cost of this between owner and occupier depends on the leases in any particular country. Other aspects, such as technological obsolescence, are unlikely to be predictable. The level of depreciation affects d in Equation 8.1, but its predictability affects the risk premium. Thus, depreciation has a dual impact on the capitalisation rate.

2. *Curable*: Some aspects are curable, albeit at a cost. Examples include building deterioration and replacement of plant. Others, such as structural specification, may not be, or may be only at a prohibitive cost. Most site factors cannot be cured by expenditure; others, such as contamination or ground conditions, may only be curable at a prohibitive cost.

3. *Diversification*: Until now, the discussion has been at the individual property level. However, as suggested in 1 above, some aspects of depreciation are related to risk, so it is possible to diversify these within a portfolio. Those affecting a particular property, a type of property or a location can be diversified. For example, technological change affects different types of properties in different ways, so its impact may be diversified by holding properties of varying age and construction. In contrast, all properties are subject to wear and tear which cannot be diversified.

4. *Relative and absolute utility changes*: Deterioration (whether to building or site) results in a change in the *absolute* utility of a property and is *internal* to the property. In contrast, obsolescence (again whether to building or site) causes changes in *relative* utility, that is, relative to properties meeting any new required specifications. It is a result of changes which are *external* to the property, such as building technology or user requirements. Depreciation is the combined effect of these two. This distinction is discussed more fully in the next section.

So far, depreciation has been defined according to its impact and has been classified according to its causes. To be useful in investment analysis, it is important to be able to measure its impact. This is now considered.

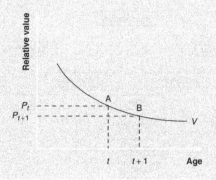

Figure 8.4▶ The impact of deterioration

8.4 Measuring depreciation

Absolute and relative price movements

As suggested in the previous section, when considering the measurement of depreciation, it is important to be aware of two separate factors which lead to price movements. Flanagan *et al.* (1989) distinguish between *absolute* and *relative* loss of utility of a capital good. When applying this analysis to property depreciation, there is a danger of confusion in the terminology.

Recall that, when considering changes in value resulting from depreciation, it is necessary to control for those which result from inflation or a general change in demand and supply. This is done by considering price movements relative to an 'equivalent new property', which also has a time-varying value. Thus, both absolute and relative loss in utility of a property can be measured with reference to its price relative to a new property.

As a property ages and suffers deterioration, its utility falls in absolute terms. This results in a change in value but note that, even if the price of the reference new property did not change, the subject property would still fall in value. This can be viewed as a *move* to the right along the age/value curve (V) as shown in Figure 8.4.

In contrast, obsolescence results in a relative loss of utility, that is, there is a *shift* in the position of curve V. A shift to the right means that for all ages the property is more valuable (for example, through site appreciation); a shift to the left means that the property is less valuable. This is shown in Figure 8.5. At any point t, a move from A on curve V to A_1 on curve V' represents appreciation from P_t to P_t'; and a move from A to A_2 on curve V'' represents depreciation from P_t to P_t''.

The combined effect of the move and shift is shown in Figure 8.6.

The time pattern of depreciation

For any property, decreases in price resulting from deterioration (movement along a curve) are likely to be reasonably constant; while those resulting from obsolescence are likely to be small for many years and then have a substantial

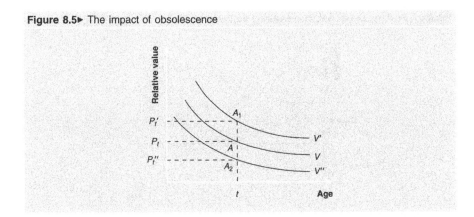

Figure 8.5► The impact of obsolescence

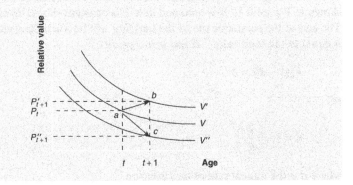

Figure 8.6► Combining the impacts of deterioration and obsolescence

effect over a period of a few years on buildings of a particular age and construction. Thus, the time pattern of price relative to an equivalent new property – that is, depreciation – is unlikely to follow a simple geometric pattern as in Figure 8.1. Therefore, while it may be reasonable to apply a constant rate at the portfolio level, this is unlikely to hold at the individual property level. This is an important issue when seeking to measure depreciation. A related issue is the affect of 'vintage', which means that, in any one year, properties of different ages and construction may be subject to very different rates of depreciation.

Before considering the two broad approaches to measuring depreciation, a simple theoretical model is discussed. This allows a broad order of magnitude of depreciation to be estimated.

A simple model

This approach does not attempt to estimate depreciation from data for actual properties but, rather, depends on a number of simple assumptions. It considers geometric depreciation in capital value and is derived from an accountancy procedure for estimating asset values (see Section 8.2 and Figures 8.1 and 8.2). When applying this approach to property, the value of a new property (V_0) is split

Figure 8.7▶ A simple model for measuring depreciation

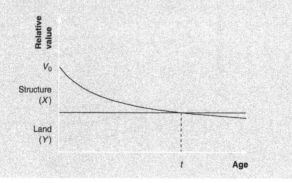

into a building or structure component (X) and a land component $(Y = V_0 - X)$ as shown in Figure 8.7.[4] It is assumed that Y is constant in real terms through time. The end of the economic life of the building will be when the total property value is equal to the land value.[5] If this is in t years:

$$V_0(1 - d)^t = Y$$

◀Equation 8.4

so:

$$d = 1 - \left(\frac{Y}{V_0}\right)^{1/t}$$

◀Equation 8.5

where d is the annual rate of depreciation.

Estimates of d for different values of Y/V_0 and t are shown in Example 8.3. The analysis has two merits. First, it permits analysis of the effect of time and the land/total value ratio on depreciation. It is clear that depreciation falls as the land component rises and as the economic life increases. Second, it enables an estimate of depreciation within a reasonable range: the results below suggest a range for the vast majority of properties of between 1.0% and 4.5%.

Consider the effect of land value and age on the rate of depreciation (see Table 8.3).

Table 8.3▶ Depreciation (%) for different values of t and Y/V_0

Land value (%)	Years								
	20	30	40	50	60	70	80	90	100
5	13.9	9.5	7.2	5.8	4.9	4.2	3.7	3.3	3.0
10	10.9	7.4	5.6	4.5	3.8	3.2	2.8	2.5	2.3
20	7.7	5.2	3.9	3.2	2.6	2.3	2.0	1.8	1.6
30	5.8	3.9	3.0	2.4	2.0	1.7	1.5	1.3	1.2
40	4.5	3.0	2.3	1.8	1.5	1.3	1.1	1.0	0.9
50	3.4	2.3	1.7	1.4	1.1	1.0	0.9	0.8	0.7
60	2.5	1.7	1.3	1.0	0.8	0.7	0.6	0.6	0.5

Table 8.4▶ Typical economic lives of income-producing properties in the USA

Type of property	Economic life (years)
Hotels and motels	30–40
Apartments	40–60
Warehouses	50–75
Banks	40–50
Offices	20–60
Stores	30–50
Shopping centres	15–30
Restaurants	10–20
Garages and repair shops	40–60
Specialty properties	15–75

Source: Corgel *et al.* (1998: 424).

This method uses age as a proxy for both building deterioration and building obsolescence for, as a building ages, it suffers from both. Further, it is assumed that there is no site depreciation (or appreciation). This is unlikely to be a reasonable assumption, particularly for the retail sector. In cases where there is systematic shortage of suitable land, and real land values have been rising, it is likely to overestimate depreciation. The basic method can be developed to incorporate growth in the land value component (see Bowie, 1983).

While the method allows a reasonable range for the *average* value to be established, a more precise estimate cannot be made because the economic life of a building is impossible to establish accurately in advance. For the USA, estimates of the typical economic lives of income-producing properties are reported by Corgel *et al.* (1998: 424) and are shown in Table 8.4.

Two other approaches have been developed to estimate depreciation. These are *time series* or longitudinal analysis and *cross-sectional* analysis. They have been used to estimate depreciation in rental value, yields (capitalisation rates) and capital value.

Time series analysis

Time series analysis uses data for one or more properties over time, that is, as they age. The longer the period of analysis, the better are likely to be the results. Ideally, properties of different vintages should be included in the analysis to determine their different time patterns of depreciation.

The time series data are affected by inflation and general market movements, so removing these by defining the time-varying prime reference value for calculating depreciation, is an important feature of this method. The required inputs are time series data on the subject property (or an index of properties) and on an equivalent new property (or an index of new properties). In practice, the latter requirement is for the prime rent in the appropriate sub-market.

Example 8.4 illustrates the estimation of rental depreciation. The first stage is to calculate rental growth each year for the subject and reference (prime) properties or indices:

$$\%\Delta R_{S_t} = \frac{100(R_{S_t} - R_{S_{t-1}})}{R_{S_{t-1}}}$$

◄Equation 8.6a

$$\%\Delta R_{R_t} = \frac{100(R_{R_t} - R_{R_{t-1}})}{R_{R_{t-1}}}$$

◄Equation 8.6b

where R_{S_t} and R_{R_t} are the rental indices at time t, respectively, for the subject properties and the reference properties.

The compound depreciation at the end of period t (assuming both indices are based on the same year) is given by:

$$D_t = \frac{R_{R_t} - R_{S_t}}{R_{R_t}}$$

◄Equation 8.7

where D_t is the compound depreciation to the end of period t.

To calculate the annual rate for period t, it is necessary to remove the depreciation of the previous periods:

$$d_t = 1 - \frac{1 - D_t}{1 - D_{t-1}}$$

◄Equation 8.8

The geometric average is calculated from:

$$d = 1 - (1 - D_t)^{1/t}$$

◄Equation 8.9

where d is the geometric average.[6]

Table 8.5► Estimating depreciation using time series data

Year	Reference		Subject		Depreciation	
	Rent index	Rental change	Rent index	Rental change	Cumulative	Annual
0	100.0		100.0		0	
1	105.0	5.0%	103.0	3.0%	1.9%	1.9%
2	111.0	5.7%	107.0	3.9%	3.6%	1.7%
3	120.0	8.1%	113.0	5.6%	5.8%	2.3%
4	150.0	25.0%	142.0	25.7%	5.3%	−0.5%
5	173.0	15.3%	164.0	15.5%	5.2%	−0.1%
6	156.0	−9.8%	140.0	−14.6%	10.3%	5.3%
7	152.0	−2.6%	132.0	−5.7%	13.2%	3.2%
8	160.0	5.3%	136.0	3.0%	15.0%	2.1%
9	166.0	3.8%	139.0	2.2%	16.3%	1.5%
10	180.0	8.4%	147.0	5.8%	18.3%	2.5%
			Average depreciation	2.0%		

As with the simple approach, this method uses age as a proxy for building deterioration and building obsolescence. Site depreciation could be considered by examining identical properties in different locations but data availability is unlikely to allow this. Data for actual properties and valuers' estimates for hypothetical buildings with specified characteristics can both be used. In practice, data limitations mean that this method is usually used in conjunction with a cross-sectional analysis.

The method is used for the City of London office market by Baum (1991) for the period 1980–6 and Barras and Clark (1996) for the period 1981–93. The former used estimates of rent and capital value made by a panel of experts, while the latter used data from the Investment Property Databank (IPD). Baum (1991) also considers industrials in Slough, to the west of London. Both use time series analysis as an adjunct to cross-sectional analysis.

Simple cross-sectional analysis

A cross-sectional study involves looking at buildings of different ages at the same time. Thus, data are collected on rental values of a new property, a one-year-old property; a two-year-old property; and so on. In effect, this approach uses buildings of different ages as proxies for a particular building(s) considered at different times. There are two different forms of the cross-sectional approach: averaging and regression.

The first calculates average values for properties of each age and uses a similar approach to the time series method outlined above. Either actual or hypothetical data may be used.

Example 8.5 illustrates the calculation of depreciation using the basic cross-sectional approach. The figures in the second column of Table 8.6 represent the rental value for properties of the given age. The third column shows the average rate of depreciation during each time period. As:

$$R_t = R_{t-5}(1 - d)^5$$ ◄Equation 8.10

then:

$$d = 1 - \left(\frac{R_t}{R_{t-5}}\right)^{1/5}$$ ◄Equation 8.11

where: R_t is the rent index value for a property of age t
R_{t-5} is the rent index value for a property five years newer
d is the average rate of depreciation during the period $t - 5$ to t.

If annual data were available, annual rates could be calculated.

Table 8.6► Estimating depreciation using cross-sectional data

Age	Rent index	Depreciation
0	100	
5	85	3.2%
10	70	3.8%
15	60	3.0%
20	50	3.6%
25	35	6.9%

Example 8.5

However, the values of depreciation estimated using this method apply only to the year of calculation. The fundamental problem with this approach is that the estimate of d is specific to the cross-section year. There are two reasons: market conditions and vintage. In a buoyant market, shortage of space may make older buildings relatively more attractive. In contrast, in a weak market, demand for poorer quality space may decline relatively more greatly. Further, buildings of one vintage may be better able to be adapted to changes in user requirements than those of another. Alternatively, the rate of change of user requirements or construction innovation may vary over time. Thus, in Example 8.5, a ten-year-old building depreciated at an annual rate of 3.5% during the previous five years but, in future, ten-year-old buildings may depreciate at different rates. Note that, for these reasons, it is not possible to estimate the time pattern of depreciation using this method.

This is the approach adopted by Salway (1986) who used hypothetical buildings in different parts of the UK. Valuers were asked for their views on the rental value and yield of the buildings if they were of a certain age. He estimated an average figure for both the office and industrial markets of approximately 3% annually. However, the buildings were identical apart from age: all the other variables (design, site value, location, quality and so on) were removed. This means that obsolescence and site depreciation were excluded. Barras and Clark (1996) also use the method but with IPD data for actual buildings in the City of London.

Cross-sectional regressions with explanatory variables

A second variant of the cross-sectional approach uses data on individual properties in a regression analysis. Depreciation is the dependent variable; and age and variables representing building design and quality and location can be used as independent variables. Thus, at any given time:

$$D_i = \alpha + \Sigma \, \beta_j x_{ij} + e_i \qquad \qquad \blacktriangleleft\text{Equation 8.12}$$

where: D_i is the total depreciation for building i
x_{ij} is the value of independent variable j for building i
α and β_j are constants specific to the cross-section
e_i is a random error.[7]

This is, in essence a hedonic model (see Chapter 4) and allows an assessment of the relative importance of factors which cause depreciation.[8] Three studies have adopted this approach (Baum, 1991; Khalid, 1992; and Yusof, 1999).

Baum (1991) considers rent, yield and capital value of offices within a small area of the City of London. He adjusts the rents to control for differences in site and location, and the yields to control for lot size. He asked one panel of valuers to assess the rents and yields of the properties and another to provide scores on a number of design and quality features for each property. He uses four composite factors representing configuration, internal specification, external appearance and physical deterioration. The shortfall below prime value was calculated as a function of the shortfall in the quality factors. He concludes that building quality is more important than age in determining depreciation. He uses a similar approach, but with rental valuation data, for industrial property in Slough, west of London. One problem of this approach is the inclusion in the composite factors of individual variables which are highly correlated, both within and between factors. This is known as *collinearity* and creates problems when interpreting the results.

Khalid (1992) adopts a similar approach to Baum but uses actual rental data for offices in Kuala Lumpur, Malaysia. In an attempt to overcome collinearity, he combines highly correlated variables into composite factors by multiplying individual variable values. The difference between the rent of a new building and that of a property of a given age is considered to be a linear function of variables: age, location (based on a survey of tenants' preferences) and the composite factors representing flexibility, quality of the building engineering services and appearance. He concludes that flexibility and appearance are most important.

Yusof (1999) also considers the Kuala Lumpur office market and develops the approach of Baum and Khalid. Before estimating hedonic models for rents and yields, she uses a method known as Principal Component Analysis (PCA). This converts correlated variables into uncorrelated (or orthogonal) factors. These are constructed to determine the most important dimensions in the original data and enable a large number of variables to be converted into a much smaller number of important factors. She reduces 37 variables (measuring the location, quality and services of the offices) to eight factors for rents and six for yields. This overcomes the serious problem of multi-collinearity between age and many of the quality variables. She is then able to relate these factors to the different causes of depreciation. However, although building depreciation and obsolescence are conceptually separate, she is unable to construct factors which measure them separately as both are functions of age and quality of construction. She finds that building quality factors are most important in explaining depreciation but some locations can mitigate its effect.

A fundamental problem with the cross-sectional approach is that the estimate of depreciation is specific to the cross-section year. Other problems apply to both time series and cross-sectional approaches. The main one is survival bias. When considering actual properties it is only possible to consider those properties which have survived to the time of the analysis. This means that buildings which have suffered substantial depreciation and have been demolished are excluded and, particularly for older properties, the values of depreciation calculated may be underestimates.

A further problem is the reliance on valuation estimates, particularly for yields and capital value, in the absence of sufficiently representative market data. As has been suggested in Chapters 3, 4 and 5, this can lead to problems of reliability. Finally, there is a need to factor out the affects of expenditure, whether regular maintenance or periodic refurbishments. These both enhance the value of a property but at a cost. In practice, the former tends to be ignored unless net rent information is available. The latter can be dealt with either by removing from the analysis properties which have undergone major refurbishment or, as Baum (1991) does, by treating them separately.

The various estimates of depreciation are summarised in Table 8.7. It can be seen that there is a reasonable degree of consistency.

Table 8.7▶ Estimates of depreciation

Source	Variable	Method	Property type	Estimated annual depreciation (%)
Table 8.3	Capital value	Simple model	All	1–4.5
Salway (1986)	Rent	CS	Hypothetical City of London offices	1.4
			Hypothetical West End of London offices	3.0
			Hypothetical UK offices	3.0
			Hypothetical UK industrials	3.2
	Capital value	CS	Hypothetical City of London offices	2.4
			Hypothetical West End of London offices	3.8
			Hypothetical UK offices	5.5
			Hypothetical UK industrials	5.3
Barras and Clark (1996)	Rent	TS	City of London offices	1.0
		CS 1980	City of London offices	1.2
		CS 1989	City of London offices	1.5
		CS 1993	City of London offices	1.2
	Capital value	TS	City of London offices	1.5
		CS 1980	City of London offices	1.8
		CS 1989	City of London offices	1.75
		CS 1993	City of London offices	1.45

Baum (1991)	Rent	CS 1986	City of London offices	1.1
		CS 1986	Slough industrials	0.52
		TS 1980–86	City of London offices	0.78
		TS 1980–86	Slough industrials	2.86
	Capital value	CS 1986	City of London offices	1.6
Lambert Smith Hampton and HRES (1997) (this report is an update of the work of Baum (1991))	Rent	CS 1996	City of London offices	2.2
			West End of London offices	1.6
	Capital value	CS 1996	City of London offices	2.9
			West End of London offices	2.2
Khalid (1992)	Rent	CS 1990	Central Kuala Lumpur offices	2.4
Yusof (1999)	Rent	CS 1997	Central Kuala Lumpur offices: traditional	23.8 (total)
			Central Kuala Lumpur offices: transitional	17.5 (total)
			Central Kuala Lumpur offices: modern	7.0 (total)

Note: The estimates vary depending on the length of the time series and the ages in the cross-section, emphasising the time varying pattern of depreciation.

Depreciation is important because it affects the return on an investment. It does this through lower rents and higher capitalisation rates, when compared to a new property, and the need for expenditure on regular maintenance and periodic refurbishment. The effect on yield is twofold: first, on the amount of depreciation, d in Equation 8.1; and, second, on the risk involved in estimating future rates of depreciation which affects the risk premium.

The traditional approach to valuation considers depreciation implicitly as part of the all-risks yield or capitalisation rate. This failure to consider depreciation explicitly can result in it being inadequately taken into account. Alternatively, cash flow models take it into account explicitly.

Depreciation is an effect which can be observed by loss in rental and capital value and by removal of properties from the market stock. The two are closely related and consideration in this chapter focused on the former. To control for general price inflation and market movements, depreciation in rent and yields is defined relative to prime property in the appropriate sub-market. The definitions of both the appropriate reference property and the appropriate sub-market are not without problems. It is important to be aware that the prime property changes through time and will have a higher specification compared to a subject property.

Depreciation can be divided into three broad categories: deterioration caused by use and environmental factors; building obsolescence caused by changing user requirements and building technology; and site depreciation, mainly the result of changes in the location requirements of users. The third of these may be positive and so compensate for loss in the value of the physical structure. While building deterioration is likely to occur at a reasonably steady rate, building obsolescence is likely to have a variable impact from period to period. Different vintages of property are likely to be differentially affected by changing user needs and technological developments.

The measurement of depreciation poses a number of problems. There are four approaches. The first is a simple model which separates the building value from the site value. The latter is assumed to be constant in real terms, so depreciation can be calculated from the economic life of the property – that is, the time until the property value equals the site value. Although the method relies on knowing the life of a property, it can be used to establish an appropriate range of values for depreciation.

The second approach, known as time series analysis, considers the values of subject properties through time compared to new prime property in the same sub-market. This combines building deterioration and obsolescence. In theory, it can be used to consider site depreciation if properties of the same type are considered in different locations, but the data availability is likely to restrict such analyses.

The third approach is simple cross-sectional analysis which considers only the age of the property. Properties of different ages in the one year are used as proxies for ageing properties. Both market conditions and the vintage of the property will affect the cross-sectional results, so it is difficult to generalise from them. Here again, age proxies for building deterioration and obsolescence. Site depreciation could be taken into account by considering identical properties on different sites, but such an approach is restricted by data availability.

The final approach is cross-sectional regression analysis with depreciation as the dependent variable and independent variables representing factors likely to cause depreciation. This is a form of hedonic model and it allows the relative importance of the factors to be assessed. In this approach, collinearity of the independent variables is a problem although it can be overcome using statistical techniques such as Principal Component Analysis.

Two problems which affect most of these approaches are the need to use valuation data and survivor bias. Valuations are estimates of market values rather than actual values and so are prone to error. Survival bias means that information is only available for properties that have not been demolished. This may understate the impact of depreciation.

Although the methods outlined in this chapter allow depreciation to be approximated, they cannot provide precise figures. This is no different from other inputs into a pricing framework. As with all inputs, it is more appropriate to consider a range of possible values rather than a single value.

The next chapter moves from the pricing framework to consideration of the practical management of a property portfolio.

Notes

1. The US literature on depreciation and property (for example, Hendershott and Ling, 1984a and 1984b; Sirmans, 1980) tends to consider taxation issues.
2. In practice, refurbishment may be difficult if the tenant is reluctant to permit it because of the impact on business. This may mean that it can only take place at the end of a lease.
3. Baum (1988, 1991) also distinguishes between tenure and property factors. For example, if the term of the lease were shortened, then the value of the lease would fall, so the property would suffer from depreciation. These tenure aspects are independent of the physical building or its location.
4. Salway (1986) estimates land value as a percentage of new property value to range from 5% for UK northern industrials to 50% for Central London offices.
5. Strictly, the cost of demolition should be taken into account but is ignored in this simple illustration.
6. A log regression could also be used to estimate the average geometric rate. This would have the advantage of including all annual points rather than only the first and last. More complex time patterns could also be fitted.
7. The precise form of the dependent variable may vary. Baum (1991) and Khalid (1992) use the *amount* of the difference while Yusof (1999) uses the total *rate* of depreciation. This has consequences for the functional form of the relationship and tests are required to ensure that the models conform to standard regression assumptions. The independent variables vary from study to study.
8. Straightforward hedonic models of offices have been calculated by Dunse and Jones (1998) and Hough and Kratz (1983). Note that the approach is unsuitable for time series analysis as the hedonic prices are likely to be time-varying.

Further reading

Barras, R. and Clark, P. (1996) Obsolescence and performance in the Central London market, *Journal of Property Valuation and Investment*, 14(4), 63–78.

Baum, A. E. (1988) Depreciation and property investment appraisal, in MacLeary and Nanthakumaran (Eds), *Property investment theory*, E & FN Spon, London, 48–69.

Baum, A. E. (1991) *Property investment depreciation and obsolescence*, Routledge, London.

Bottum, M. S. (1988) Estimating economic obsolescence in supply saturated office markets, *The Appraisal Journal*, October.

Bowie, N. (1983) The depreciation of buildings, *Journal of Valuation*, 2(1), 5–13.

Boykin, J. H. and Ring, A. A. (1993) *The valuation of real estate*, Regents/Prentice Hall, Englewood Cliffs (fourth edition): Chapter 12.

Brown, G. (1985) A note on the analysis of depreciation and obsolescence, *Journal of Valuation*, 4(3), 230–8.

Cantwell, R. C. (1988) Curable functional obsolescence: deficiency requiring substitution or modernisation, *The Appraisal Journal*, July.

Debenham, Tewson and Chinnocks (1985) *Obsolescence: its effect on the valuation of property investments*, DTC, London.

Dixon, T. J., Crosby, N. and Law, V. K. (1999) A critical review of methodologies for measuring rental depreciation applied to UK commercial real estate, *Journal of Property Research*, 16(2), 153–80.

Dunse, N. and Jones, C. (1998) A hedonic price model of office rents, *Journal of Property Valuation and Investment*, 16(3), 297–312.

Harps, W. S. (1990) Depreciated cost approach to value, in *The real estate handbook*, Seldin, M. and Boykin, J. H. (Eds), Dow Jones-Irwin, Homewood (IL) (second edition), 468–91: Chapter 28.

Hough, D. E. and Kratz, C. G. (1983) Can good architecture meet the market test?, *Journal of Urban Economics*, 14, 40–54.

Jones Lang Wootton (1987) *Obsolescence: the financial impact on property performance*, JLW, London.

Khalid, A. G. (1992) *Hedonic price estimation of the financial impact of obsolescence on commercial office buildings*, Unpublished PhD thesis, University of Reading.

Lambert Smith Hampton and HRES (1997) *Trophy or tombstone? A decade of depreciation in the Central London office market*, Lambert Smith Hampton and HRES, London.

Malpezzi, S., Ozanne, L. and Thibodeau, T. G. (1987) Microeconomic estimates of housing depreciation, *Land Economics*, 63(4), 372–85.

Miles, J. (1986) Depreciation and valuation accuracy, *Journal of Valuation*, 5(2), 125–37.

Salway, F. (1986) *Depreciation of commercial property*, CALUS, University of Reading.

Sykes, S. (1984) Periodic refurbishment and rental value growth, *Journal of Valuation*, 3(1), 32–41.

Yusof, A. (1999) *Modelling the impact of depreciation: a hedonic analysis of offices in the City of Kuala Lumpur, Malaysia*, Unpublished PhD thesis, University of Aberdeen.

Chapter 9

Constructing and managing a property portfolio

9.1 Introduction

This chapter applies much of the material presented in previous chapters to the practical construction and management of property portfolios. Traditionally, property investment focused on individual buildings, and a property portfolio was seen as a collection of buildings rather than an aggregate investment comprising particular broad categories, with its own structure and characteristics. Analyses of the property market took little explicit consideration of the economic context. Since the mid-1980s, the focus has shifted significantly to the strategic level and the property market is now viewed in the context of the other capital markets and the wider economy. The reasons for the expansion of this approach include: pressure from large investors and (in the UK) from advising actuaries; the integration of property into strategies for mixed-asset portfolios; pressure from, and developments in, other investment markets; and the wider availability of the data and techniques necessary to undertake the analysis. The strategic approach is now widely accepted and trustees of many pension funds now expect it. Nonetheless, in property, as in other investment markets, some fund managers still focus on the selection of individual buildings (the stock).

The approach outlined in this chapter is called *active* portfolio management and contrasts with *passive* management. Passive portfolio management involves setting up a portfolio with particular characteristics, such as matching an index, and trading only to maintain these characteristics.[1] A fund which exactly replicates an index is called an *index fund*. Such a fund has the advantage of low costs as it requires only a minimal amount of management, research and dealing. An index fund is guaranteed market performance, less management costs. The approach is popular in the share market where it is possible to replicate an index exactly and so to guarantee market performance, less costs.

Index funds have developed from a view that it is not possible to beat a market or competitor benchmark consistently, at least not without taking higher risk. If risk-adjusted returns are considered, this is not to suggest that some funds will not beat the benchmark: indeed, if fund performance were symmetrically distributed, half of the funds would underperform and half would overperform each year. However, the *expected* (risk-adjusted) outperformance is zero and, over time, the average outperformance of any one fund will approach zero. A refinement suggests that it may be possible to outperform but the cost of active management cancels any outperformance. These views are consistent with the *Efficient Markets Hypothesis* (see Chapter 5).

There is a fundamental problem in constructing an index fund in the property market because no two properties are identical, so no fund could hold the same properties as the index. There remains the risk of performance different from the market because of the specific risk of the properties in the portfolio, and large property lot size means that it is difficult to diversify away this risk (see Chapters 2, 4 and 7). For a small fund, the specific risk may be significant (see below) and active asset management is required of the buildings, regardless of any benchmarking strategy.

The contrary view to passive management is that active management can improve performance. Its purpose is to increase performance and to control risk. Central to active management is an assumption that the market contains inefficiencies which can be exploited by superior research (see Chapter 5). It is impossible to produce conclusive evidence to support active management as it would take many years for the outperformance to be statistically significant.[2] However, whatever the arguments in favour of the two approaches, active management dominates all investment markets. Section 9.2 presents an overview of active portfolio management and applies it to property portfolios. Section 9.3 considers structure-related issues in active property portfolio management, while Section 9.4 examines stock (individual buildings) related issues and other practicalities. The issues are illustrated by a notional portfolio. Section 9.5 provides a summary and conclusions.

9.2 Approaches to active fund management

Before examining property fund management in Section 9.3, this section considers some more general aspects of active fund management. It first considers the different dimensions of active management, including structure, stock selection and asset management. The consequence for portfolio strategy of the availability of these skills is then examined. Finally, the process of setting a strategy is compared with a business plan and the stages are set out for a property portfolio.

Adding value

In general, the purpose of active portfolio management is to achieve above market performance by careful risk management. This is sometimes called *adding value* to a portfolio. For share portfolios, there are two ways to do this: setting the structure and stock (individual shares) selection. For a property portfolio, there is also active asset management of the stock (individual buildings) by, for example, refurbishments and developments.

Setting the structure involves determining how much to invest in each broad category. The conventional divisions of a share portfolio in the UK are the industrial sectors of the FT-Actuaries All-Share Index. In the property market it is conventional to use market sectors, such as offices, retail, industrials and housing along with geographical regions.[3] As for forecasting (see Chapter 6), this level of spatial aggregation is a convenience and there is evidence to suggest that other classifications may be more appropriate (see Hoesli, Lizieri and MacGregor, 1997, for a review of the literature).

The structure is set in two stages: first choosing a long-term structure, the benchmark; and second, taking short-term positions around that benchmark. Decisions are not taken in isolation from competitors, as to do so could create competitor risk (see Chapters 3, 7 and 10, and below). Funds are likely to be attractive to investors if they have above-average performance, and fund managers are likely to get business if they have above-average performance.[4] 'Average' in this sense means when compared to competitors, thus, a neutral or long-term position is defined by a benchmark of competitors.[5] In practice, as it may be difficult to obtain detailed information on historical performance and the detailed structure of competitor funds, a market index is used as a substitute.

Short-term positions or 'bets' are taken relative to the benchmark on the basis of forecasts of return in the individual asset markets. Thus, for example, if the share market were expected to do well, a long position would be taken, that is, a greater proportion than the benchmark. If a market were expected to do badly, a short position would be taken.

As well as expected return, risk is also an important consideration in investment strategy. To obtain above-average return means taking bets (long or short positions) relative to the market and for these bets to be correct. There is always the risk that the forecasts will be wrong and the positions will produce performance lower than the market. This risk, rather than the traditional measure of volatility of expected returns, is a central feature of investment management in practice. It is measured by the *tracking error* which is the standard deviation of the difference between the fund's expected return and that of the market.[6]

The structural bets form the basis for stock (buildings) selection and asset management. The former identifies the types of shares or buildings to be bought and sold, but not the specific buildings. It involves choosing the individual shares or properties and requires detailed knowledge of these. The objective is to buy buildings with above-average expected performance. The smaller the fund, and so the greater the specific risk, the more important is stock (building) selection. While specific risk is easily diversified in a share portfolio, it remains an important feature in many property portfolios (see Chapters 4 and 7 and below).

For a property portfolio, value can also be added through asset management of the buildings in the portfolio, including refurbishment, redevelopment or changes to lease structures. It involves assessing potential and planning the necessary work, in a way which fits into the fund's requirements and its views on sub-markets, to maximise returns.

Figure 9.1▸ Property fund management strategies

		Manager's ability at setting structure	
		Good	**Bad**
	Good	**Modern management** Take positions based on forecasts. Use pricing models to identify buildings to buy and sell.	**Traditional management** Take benchmark positions. Use pricing models to identify buildings to buy and sell.
Manager's ability at stock selection	**Bad**	**Large funds** Take positions based on forecasts. Diversify by buying a lot of properties	**Passive management** Take benchmark positions. Diversify by buying a lot of properties.

Source: Adapted from Baum and Lee (1990).

Portfolio management skills

The skills required for setting a structure and for stock selection are quite different. For a property portfolio, the former requires the type of analyses covered in this book; while the latter requires more traditional surveying skills, although, given the predominance of conventional implicit valuation methods, there is clearly room for improvement (see Chapters 2 and 5). The appropriate strategies for managing a property fund can be classified according to the skills of the managers (see Figure 9.1):

1. A fund management organisation able to combine both structure setting and stock selection skills should seek to add value by taking positions relative to the benchmark and by stock selection and asset management using pricing models to identify buildings to buy and sell. This may be termed *modern management*.
2. Skills only in setting structure mean taking positions but otherwise buying a large number of buildings to diversify specific risk. This is known as naïve diversification and may be a preferred strategy for *large funds* which cannot reasonably expect consistently to select properties with above-average expected returns.
3. Skills only in stock selection require benchmark weights and a focus on mispriced properties. This focus on stock selection may be termed a *traditional management* approach. Until the later 1980s, property portfolio management in the UK was dominated by surveyors who concentrated on stock selection, with little explicit analysis of the portfolio structure or the benchmark. As property fund management has been integrated more with other investment markets and as other skills have been applied to property fund management, the focus has shifted towards the two left-hand quadrants.

4. Skill in neither demands a *passive management* strategy of benchmark weights and naïve diversification. This is the property market equivalent of an index fund but it retains tracking error risk as each building is unique, and asset management is still required of the stock.

Devising a modern management strategy for a property portfolio is rather like producing a business plan. It should contain:

1. A clear statement of *objectives*: the benchmark, return, risk and timescale.
2. *Portfolio analysis*: the current structure relative to the benchmark; and an analysis and explanation of recent performance.
3. *Forecasts*: market return forecasts; and calculation of fund expected return and risk attributable to structure.
4. Calculation, as appropriate, of *expected stock return and risk*.
5. Consideration of *achievement of objectives* if no action is taken.
6. Consideration of the constraints on restructuring arising from *stock characteristics*.
7. Consideration of *other practicalities*.
8. A *strategy* for the portfolio: proposed new structure; identification of which buildings to sell; identification of the type and location of buildings to be bought; and identification of required asset management.

Number (1) sets out what the fund strategy is trying to achieve; (2) shows where it is now and explains how it got there and what it was good and bad at; (3) to (7) assess the opportunities and constraints on achieving the objectives; and (8) provides the strategy for achieving the objectives.

These issues are considered in the next two sections and are illustrated by a notional portfolio. First, predominantly structure issues are considered and then the focus shifts to stock issues.

9.3 Structure considerations

In this and the next section, a notional portfolio will be used to illustrate the analysis of a property portfolio and the development of an appropriate strategy. The Management Employees in Surveying Superannuation (MESS) Fund is an immature pension fund with £100m in its property portfolio. It has performed poorly throughout the 1990s and new managers have been appointed to attempt to improve performance.

Objectives

Typically the property portfolio is one part of a larger, mixed-asset fund. It is highly unlikely there will be any consideration of liabilities (see Chapters 2, 3 and 9) at the property level. The mixed-asset fund strategy might determine whether the property fund is able to take high or low risks, and so influence its objectives, but even this is unlikely to be explicit. More usually there is no statement of the purpose of the property component of the fund and the objectives would be determined by changing market practice and convention.

The benchmark

It is unlikely that a fund would want to make decisions in isolation from its competitors as to do so would create tracking error risk. Accordingly, the return and risk objectives are defined relative to a benchmark. Timing is also important. Thus, there are four types of objective:

- the benchmark against which to measure fund performance
- the return objective relative to the benchmark
- the risk tolerance of the fund relative to the benchmark
- the timescale over which the objectives are to be achieved and so performance should be measured.

An appropriate benchmark is one which contains funds of a similar size and in a similar business area. Possible categories are large life funds, large occupational pension funds, small unitised funds and so on. These benchmarks, or other customised ones, are available in the UK and in other countries from the Investment Property Databank (IPD) (see Chapter 4). The IPD is by far the largest property index in the UK and comprises the majority of the institutional property investment market. It provides detailed performance analyses of funds.

In practice, many fund trustees select the IPD annual (all funds) index as a benchmark but this can create problems. When funds are divided into four categories (insurance, pension, short term and other), the average range each year between the highest returns and the lowest returns is 5% (Investment Property Databank, 1997). Funds within each category have different average sizes and, apart from the skills of the managers in setting the structure and in stock selection, many of the reasons for these differences relate to fund size:

- the ability of larger funds to invest in large lot size markets
- the asset management opportunities in a large portfolio containing many properties
- the ability of larger funds to undertake developments
- the lesser impact of stock specific characteristics in a large portfolio because of greater opportunities to diversify
- the ability of smaller funds to change structure quickly.

Thus, for example, smaller funds are not able to invest in very high value markets so, if these do well, a small fund will do badly *relative to the market as a whole*. On the other hand, if they do badly, the fund will do well relatively. This is precisely what happened during the second half of the 1980s in the UK. For several years small funds significantly underperformed the IPD average because they were unable to gain exposure to the high-value properties in the central London office market. Then, when this market crashed worse than the property market as a whole, the small funds outperformed the IPD average. In neither case can the differential performance be attributed to the ability of the managers, emphasising the importance of selecting an appropriate benchmark.

Return, risk and timescale

Having set the benchmark, the next stage is to specify the return objective. In the past, return objectives were often of the form 'to achieve x%' or 'to achieve *y*% above conventional gilts'. While either might guide a decision to invest in property as an asset class, they are inappropriate as objectives against which to judge the abilities of fund managers as they do not take into account factors beyond the control of a manager, such as general property market conditions and gilt market returns.

Accordingly, the return objective is now usually expressed relative to the benchmark. There are two variants, one relative to the *mean* fund performance and the other relative to the *median*. Thus, the objective could be 'to achieve average 1% above the benchmark' or 'to achieve upper quartile performance'. The latter avoids the problem of a small number of funds producing very high/low performance, so that the mean is much higher/lower than the median. The problem would be more acute if the funds with unusually high or low performance were large. This again illustrates the need to choose the benchmark carefully. In subsequent analyses, a return objective relative to the mean will be used as this is easier to manipulate statistically.

Most investments carry risk and there is a trade-off between expected return and risk (see Chapter 3). Thus, the higher is the required return, the higher is the risk to be taken. Funds adopting high-risk strategies to generate high returns may produce very low returns. It is, therefore, essential to have a clear statement of the risk tolerance of a fund. This might, for example, be that a greater than one in three chance of underperforming the benchmark is unacceptable. This can be calculated from the *tracking error*, that is the standard deviation of the difference between the fund's expected return and that of the benchmark (see Chapter 7 and below).

Finally, timescale is important in two aspects of setting fund objectives. First, new managers who take over a badly performing fund will need time to implement a strategy, so objectives should reflect a progression over a number of years towards a desirable level of performance. Second, it is unrealistic to expect to achieve return objectives every year: even good managers get it wrong sometimes. It is, therefore, more realistic to set a target in terms of three- or five-year averages. This is particularly true for small funds where property-specific risk is significant.

The following analyses will be illustrated by a notional portfolio, the MESS Fund, the objectives of which are:

1. As from year 1, to achieve performance above the IPD medium sized pension funds average each year.
2. By three years, to achieve outperformance of 1%.
3. Beyond three years, to achieve a three-year rolling average outperformance of 1%.
4. After year 1, in any year, to have less than 40% probability of underperforming the benchmark.

Note that these objectives require a customised benchmark. Given the state of the fund (see below), they may be considered ambitious but they are the best any trustees are likely to agree. At least they allow a couple of years to turn the fund round and, thereafter, performance is judged on a rolling average. In practice, it is advisable that the process of determining objectives should be through a discussion between trustees and managers, combining what is possible with what is desirable.

Table 9.1► MESS Fund structure

Benchmark weights

	Retail	Offices	Industrials	Housing	Total
North	10.0%	5.0%	2.0%	3.0%	20.0%
Central	15.0%	10.0%	2.0%	3.0%	30.0%
South	15.0%	25.0%	6.0%	4.0%	50.0%
Total	40.0%	40.0%	10.0%	10.0%	100.0%

MESS Fund weights

	Retail	Offices	Industrials	Housing	Total
North	8.0%	6.0%	1.0%	2.0%	17.0%
Central	10.0%	15.0%	2.0%	2.0%	29.0%
South	12.0%	30.0%	10.0%	2.0%	54.0%
Total	30.0%	51.0%	13.0%	6.0%	100.0%

MESS Fund bets

	Retail	Offices	Industrials	Housing	Total
North	−2.0%	1.0%	−1.0%	−1.0%	−3.0%
Central	−5.0%	5.0%	0.0%	−1.0%	−1.0%
South	−3.0%	5.0%	4.0%	−2.0%	4.0%
Total	−10.0%	11.0%	3.0%	−4.0%	0.0%

Portfolio analysis

Portfolio analysis comprises consideration of the current structure relative to the benchmark and an analysis and explanation of recent performance.

Structure relative to the benchmark

Table 9.1 shows the structure of the benchmark, the structure of the fund and the bets, that is the difference between fund and benchmark weights. It can be seen that the fund has negative bets (short positions) in all of the retail and housing sub-markets, positive bets (long positions) in the office sub-markets and a mixture in the industrial sub-markets. This information will be used to develop a strategy. First, information on the fund's historical structure is used in the next stage to analyse recent performance.

Analysis of recent performance

It is possible to disaggregate the fund performance into benchmark, structure and stock components. For each year:

$$RF_T = RB + RF_{SE} + RF_{ST}$$

◄Equation 9.1

where: RF_T is the fund total return
RB is the benchmark return
RF_{SE} is the fund return attributable to structure
RF_{ST} is the fund return attributable to stock.

The fund and benchmark returns are obtained as the weighted averages of the returns in the sub-markets:

Table 9.2► The components of recent MESS Fund performance (%)

	1993	1994	1995	1996	1997
Benchmark	−1.0	22.0	10.5	2.5	11.0
Structure	−0.7	−1.2	−0.2	−0.5	−0.8
Stock	−0.8	−1.0	−1.1	−1.3	−1.2
Fund	−2.5	19.8	9.2	0.6	9.0

$$RF_T = \Sigma\, w_i RF_i$$ ◄**Equation 9.2**

$$RB = \Sigma\, W_i RB_i$$ ◄**Equation 9.3**

where: w_i is the fund weight in sub-market i
RF_i is the fund return in sub-market i
W_i is the benchmark weight in sub-market i
RB_i is the benchmark return in sub-market i.

The structure component is obtained by applying fund weights to the market returns and subtracting the market return from this.[7] The difference between the two is explained by the fund bets (the differences between fund and benchmark weights):

$$RF_{SE} = \Sigma\, w_i RB_i - \Sigma\, W_i RB_i$$
$$= \Sigma\, (w_i - W_i) RB_i$$ ◄**Equation 9.4**

The stock component is the residual difference between fund and benchmark returns:

$$RF_{ST} = RF_T - RF_{SE} - RB$$ ◄**Equation 9.5**

Table 9.2 presents the above analysis for the MESS Fund. The first row shows the benchmark returns while the final row shows the MESS Fund returns. The difference between the two is attributed to structure and stock factors in rows two and three. It is clear that the fund has been badly managed. The contributions to performance of both structure and stock have been consistently negative. The managers appear to have been poor at structure, stock selection and asset management.

For each year, this analysis can be disaggregated by sub-market. Table 9.3 shows fund and market returns and disaggregations into structure and stock components for 1997. The retail sector bet creates a substantial negative structure effect, while the office sector produces a positive effect. The other sector bets are smaller and their effect is slight. The stock effects were generally negative or zero. Although the fund's industrial properties in the North and Central regions performed particularly poorly relative to the market, the small weights in these sub-markets reduced the impact. The largest effect was in the office sector where large weights combined with poor performance.

Other types of disaggregations of return are also possible and useful. The IPD subdivides total portfolio returns into the standing investment (properties held throughout the analysis period), the impact of expenditure through asset management, the impact of transactions and the impact of developments.

Table 9.3► Disaggregations of structure and stock effects on returns for 1997

Fund returns

	Retail	Offices	Industrials	Housing
North	11.0%	8.0%	0.0%	3.0%
Central	15.0%	5.0%	0.0%	5.0%
South	15.0%	8.0%	8.0%	12.0%

Market returns

	Retail	Offices	Industrials	Housing
North	12.0%	5.0%	15.0%	5.0%
Central	15.0%	7.5%	10.0%	5.0%
South	18.0%	10.0%	5.0%	5.0%

Structure effect

	Retail	Offices	Industrials	Housing	Total
North	-0.2%	0.1%	-0.2%	-0.1%	-0.4%
Central	-0.8%	0.4%	0.0%	-0.1%	-0.4%
South	-0.5%	0.5%	0.2%	-0.1%	0.1%
Total	-1.5%	0.9%	0.1%	-0.2%	-0.8%

Stock effect

	Retail	Offices	Industrials	Housing	Total
North	-0.1%	0.2%	-0.2%	0.0%	-0.1%
Central	0.0%	-0.4%	-0.2%	0.0%	-0.6%
South	-0.4%	-0.6%	0.3%	0.1%	-0.5%
Total	-0.4%	-0.8%	-0.1%	0.1%	-1.2%

Note: Figures may not add up because of rounding.

Table 9.4► The impact of management, transactions and developments

	Average annual result 1980–96	Number of positives	Number of negatives
Asset management	-0.13%	6	10
Transactions	0.30%	16	0
Developments	-0.20%	9	7

Source: IPD (1997).

The published figures for the impact on the standing portfolio are summarised in Table 9.4 and clearly show the positive impact of transactions and the overall negative impact of asset management and development.[8] These totals, of course, contain a wide range of values for individual funds.

Forecasts

The next stage is to apply forecasts of return and risk to the portfolio structure to produce the expected return relative to the benchmark and its standard deviation, the tracking error. Together these can be used to calculate the probability of achieving the fund's objectives if no action is taken and to inform the appropriate buying and selling strategy.

Table 9.5► Market forecasts

	Retail	Offices	Industrials	Housing
North	10.0%	4.0%	4.0%	5.0%
Central	8.0%	2.0%	0.0%	4.0%
South	12.0%	−5.0%	2.0%	5.0%

Figure 9.2► MESS Fund bets and forecast returns

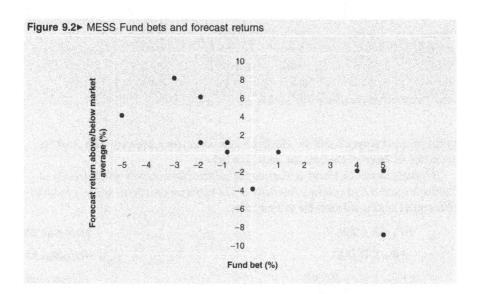

Market return forecasts

The forecasts of return can be for any appropriate period although most fund managers require annual forecasts. These are produced using the procedures outlined in Chapter 6, combining rental forecasting models with pricing models. Forecasts for next year (1998) for the benchmark of the MESS Fund are shown in Table 9.5.

Expected fund return if no action is taken

The size of the funds bets in each sub-market can be plotted against the expected returns relative to the market average expected return:

1. Positive bets in sub-markets with above-average expected return mean positive expected return relative to the market average.
2. Positive bets in sub-markets with below-average expected return mean negative expected return relative to the market average.
3. Negative bets in sub-markets with above-average expected return mean negative expected return relative to the market average.
4. Negative bets in sub-markets with below-average expected return mean positive expected return relative to the market average.

These are shown for the MESS Fund in Figure 9.2. One of the bets has a marginally positive effect (type 1 above), one is neutral and all the others have a

Table 9.6► Expected returns

Fund	2.6%
Market	3.8%
Difference	-1.2%

Table 9.7► Disaggregation of expected relative return attributable to structure

	Retail	Offices	Industrials	Housing	Total
North	-0.2%	0.0%	0.0%	-0.1%	-0.3%
Central	-0.4%	0.1%	0.0%	0.0%	-0.3%
South	-0.4%	-0.3%	0.1%	-0.1%	-0.6%
Total	-1.0%	-0.1%	0.0%	-0.2%	-1.2%

Note: Figures may not add up because of rounding.

negative effect (types 2 and 4). The figure also indicates a problem with tracking error risk as some of the bets are large (see below).

The analysis can be formalised using the information on bets and forecasts in Tables 9.1 and 9.5 to produce *forecasts* of the benchmark return, and of fund and differential return attributable to structure:

$$RF_B = \Sigma \, w_i RB_i \qquad \qquad \text{◄Equation 9.6}$$

$$RB = \Sigma \, W_i RB_i \qquad \qquad \text{◄Equation 9.7}$$

$$RF_{SE} = \Sigma \, (w_i - W_i) RB_i \qquad \qquad \text{◄Equation 9.8}$$

where: RF_B is the expected return of the fund assuming benchmark returns in each sub-market

RB is the benchmark expected return

RF_{SE} is the expected return of the fund attributable to differences between its structure and that of the benchmark.

These calculations produce expected returns as shown in Table 9.6 and the contributions of the bets to relative return are shown in Table 9.7. The expected relative return is −1.2%: the major problem lies in the retail sector and restructuring is required if the fund's objectives are to be achieved.

There is also a component of the expected return which derives from stock-specific characteristics. For all properties this specific return is constructed to have zero mean. It is reasonable to assume that large portfolios will be unable consistently to select properties with positive-specific returns, so the expected specific return can be assumed to be zero (unless there is contrary historical evidence). For small portfolios, specific return should be taken into account. To do so, it is necessary to assess the stock and to estimate specific return. This is not easy and involves an element of subjectivity. Section 9.4 below considers specific return for the MESS Fund.

Expected fund risk if no action is taken

As for return, there are two components to tracking error risk, one for the structure and one for the stock. The structure component of risk is considered

Figure 9.3► The structure of the correlation matrix

		North				Central				South			
		R	O	I	H	R	O	I	H	R	O	I	H
North	R O I H			1				2				2	
Central	R O I H			2				1				2	
South	R O I H			2				2				1	

Note: R = Retail; O = Offices; I = Industrials; H = Housing. 1 and 2 represent sub-matrices (see text).

here, while the stock component is considered in the Section 9.4. While the calculation of expected return due to structure requires trivial arithmetic, the calculation of expected structure risk is more complicated, as it requires not only the individual sub-market risks but also the correlations between the sub-markets. Historical values of the standard deviations of returns, if these exist, are of some value in assessing risk. Risk measures may also be informed by an examination of the regional industrial structures which generate demand for property: the demand for property in a well-diversified regional economy is likely to be more stable than in a region dependent on a small number of industries (see Hartzell *et al.*, 1993, for a discussion of diversification in Europe based on economic structure). Similarly, regions with economies heavily dependent on exports are likely to face greater volatility than those relying mainly on national demand. Market liquidity may also need to be taken into account.

Historical information on correlations is of some value but is generally less reliable than for standard deviations.[9] One way of overcoming the problems is to impose a simple structure on the correlation matrix. The correlation matrix can be broken into two types of symmetrical sub-matrices: one for correlations between sub-markets within the *same* region; the other for correlations between sub-markets in the *different* regions. The basic structure is shown in Figure 9.3. The national correlations, which are likely to be more reliable than regional data, can be used to inform the construction of matrix 1. Matrix 2 can then be constructed relative to it. For illustration, in this case, the inter-region matrix has been constructed using three simple rules:

1. The diagonal is based on average correlations between regions in the same sector. These are generally strong and are strongest for the retail and housing markets.
2. If the correlation is between housing and retail, the off-diagonal element is the equivalent numbers in matrix 1, less 0.1.

Table 9.8► Standard deviations and correlations

	NR	NO	NI	NH	CR	CO	CI	CH	SR	SO	SI	SH	Risk (SD)
NR	1												10%
NO	0.8	1											12%
NI	0.6	0.8	1										14%
NH	0.8	0.6	0.6	1									12%
CR	0.9	0.6	0.4	0.7	1								10%
CO	0.6	0.8	0.6	0.4	0.8	1							12%
CI	0.4	0.6	0.8	0.4	0.6	0.8	1						14%
CH	0.7	0.4	0.4	0.9	0.8	0.6	0.6	1					12%
SR	0.9	0.6	0.4	0.7	0.9	0.6	0.4	0.7	1				10%
SO	0.6	0.8	0.6	0.4	0.6	0.8	0.6	0.4	0.8	1			12%
SI	0.4	0.6	0.8	0.4	0.4	0.6	0.8	0.4	0.6	0.8	1		14%
SH	0.7	0.4	0.4	0.9	0.7	0.4	0.4	0.9	0.8	0.6	0.6	1	12%

3. Otherwise, the off-diagonal elements are the equivalent numbers in matrix 1, less 0.2.

The risk measures used in this chapter are shown in Table 9.8.
The structure component of risk is defined (see Chapter 7) as:

$$\sigma_{SE} = \sqrt{\Sigma\Sigma w_i w_j \sigma_i \sigma_j \rho_{ij}}$$ ◄**Equation 9.9**

The structure component of *tracking error risk* is defined as:

$$TE_{SE} = \sqrt{\Sigma\Sigma (w_i - W_i)(w_j - W_j)\sigma_i \sigma_j \rho_{ij}}$$ ◄**Equation 9.10**

where: σ_{SE} is the standard deviation of fund returns attributable to structure
TE_{SE} is the structure component of the tracking error
w_i is the weight of the fund in sub-market i
W_i is the weight of the benchmark in sub-market i
σ_i is the standard deviation of returns in sub-market i
ρ_{ij} is the correlation between sub-markets i and j.

Note that, if the fund and the benchmark have equal weights, the structure component of tracking error risk is zero.

Using the above formulae, for the MESS Fund, the structure risk is 9.8% and the tracking error is 1.3%. Assuming that returns are normally distributed, the probability distributions of absolute returns and of returns relative to the benchmark (both *attributable to structure*) can be calculated. The latter is of more interest and is shown in Figure 9.4. The distribution is centred on the expected relative return (−1.2%) and the tracking error (1.3%) is its standard deviation. The area under the curve to the right of a particular value gives the probability of achieving at least that relative return: sample values are shown in Table 9.9. Ignoring stock effects (which are considered in the next section), the probability of beating the benchmark is only 0.17.

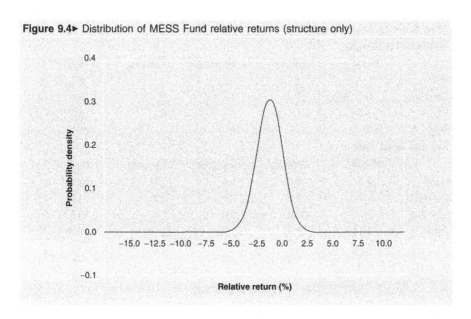

Figure 9.4► Distribution of MESS Fund relative returns (structure only)

Table 9.9► Probability of achieving at least a given relative return

Relative return (at least)	Probability
−3.0	0.92
−2.5	0.84
−2.0	0.73
−1.5	0.59
−1.2	0.50
−1.0	0.43
−0.5	0.29
0.0	0.17
0.5	0.09
1.0	0.04

Clearly, drastic changes in the structure are required if the fund is to meet its objectives. The retail bets create most of the problem for expected return, and the sizes of both retail and office bets create tracking error problems. The efficient frontier *based on structure alone* can be constructed (see Chapter 7) using expected returns relative to the benchmark and tracking error risk. For a specified return objective, the minimum tracking error risk and appropriate fund bets can be calculated. For a large fund, these provide the basis for a strategy for restructuring as both stock relative return and tracking error can be assumed to be zero. However, practicalities such as the required volume of sales (see below) must be considered. For a small fund, such as the MESS Fund, it is essential to consider the stock characteristics and the consequent stock relative return and tracking error.

Table 9.10► MESS Fund stock

Number of buildings

	Retail	Offices	Industrials	Housing	Total
North	4	1	1	1	7
Central	6	2	1	1	10
South	4	1	2	1	8
Total	14	4	4	3	25

Average value (£m)

	Retail	Offices	Industrials	Housing	Total
North	2.0	6.0	1.0	2.0	2.43
Central	1.7	7.5	2.0	2.0	2.90
South	3.0	30.0	5.0	2.0	6.75
Total	2.14	12.75	3.25	2.00	4.00

9.4 Stock and other considerations

This section considers the remaining steps in the development of a portfolio strategy. These are:

► calculation of the expected return and tracking error risk from stock
► consideration of achievement of objectives if no action is taken
► consideration of the constraints on restructuring arising from stock characteristics
► consideration of other practicalities
► a strategy for the portfolio: proposed new structure; identification of which buildings to sell; identification of the type and location of buildings to be bought; and identification of required asset management.

The traditional approach to property portfolio management was to look for 'good' buildings, however defined, regardless of the impact on structure and in the absence of an analysis of fund return and risk. However, an investment in an individual building which may appear 'good' in its own right need not be good in a portfolio context. It could, for example, substantially increase tracking error risk if it were in a sector in which the fund were already overweight. For a small fund, consideration of stock (buildings) is an important aspect of portfolio strategy and it is necessary to consider structure and stock together and iteratively. The number and value of buildings in the MESS Fund are shown in Table 9.10. For simplicity, it is assumed that buildings within each sub-market have the same value. The most obvious feature of the fund is the small number of high-value offices, particularly the large office in the South.

Stock return and risk

As discussed above, for a large portfolio, it is reasonable to assume that the expected return attributable to stock is zero; for a small portfolio, this is unlikely to be true. In the case of the MESS Fund, there is strong evidence of a consistently negative impact of stock. The average over the last five years was nearly −1.1% and over the last three years was −1.2% (see Table 9.2). If nothing is done, it is not unreasonable to suppose that this will continue (although detailed consideration of the stock may be required to confirm this, at least subjectively). The structure and stock components of expected relative return may be added. So the MESS Fund total expected relative return for next year is −2.4% (−1.2% from structure and a further −1.2% from stock).

The stock risk comes from the individual properties and depends on the number and sizes (values) of the buildings. As the risks are specific to individual buildings, they are uncorrelated, so the formula is:

$$TE_{ST} = \sqrt{\Sigma\, p_i^2 \sigma_i^2}$$

◀Equation 9.11

where: TE_{ST} is the fund tracking error attributable to stock
p_i is the proportion of the fund in building i
σ_i is the specific risk of building i.[10]

For simplicity, and in the absence of contrary evidence, it is assumed that the property specific risks are equal, thus:

$$TE_{ST} = \sqrt{\sigma^2 \Sigma\, p_i^2}$$

◀Equation 9.12

Note that the stock components of tracking error and absolute risk are identical (see Chapter 7).

For illustration, assume that all n properties in the portfolio have an equal value, then $p_i = 1/n$ for all i, and:

$$
\begin{aligned}
TE_{ST} &= \sqrt{\sigma^2 \Sigma \left(\frac{1}{n}\right)^2} \\
&= \sqrt{\sigma^2 n \left(\frac{1}{n}\right)^2} \\
&= \sqrt{\frac{\sigma^2}{n}}
\end{aligned}
$$

◀Equation 9.13

Figure 9.5 shows how dramatically TE_{ST} falls as n, the number of properties, increases (assuming a value for σ of 20).[11]

By construction, the only systematic factors are sector and region, so all remaining risk must be specific to the individual properties and is uncorrelated to the structure component of risk, thus the two variances can be added and the total tracking error is the square root:

$$TE_T = \sqrt{TE_{SE}^2 + TE_{ST}^2}$$

◀Equation 9.14

where: TE_T is the total tracking error
TE_{SE} is the structure component
TE_{ST} is the stock component.

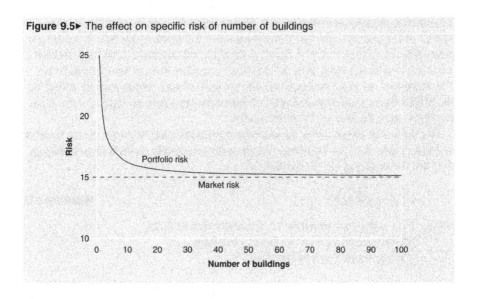

Figure 9.5► The effect on specific risk of number of buildings

Thus, even if the portfolio has an identical structure to the benchmark, and TE_{SE} is zero, there remains stock risk.

Applying this analysis to the MESS Fund produces values of 6.9% for TE_{ST} and 7.0% for TE_T. (Note that if the buildings were of equal value, TE_{ST} would be $20\%/25^{1/2} = 4\%$.)

Achievement of objectives if no action is taken

It is now possible to consider whether the fund can meet its objectives. Recall that these were:

1. As from year 1, to achieve performance above the IPD medium-sized pension funds average each year.
2. By three years, to achieve outperformance of 1%.
3. Beyond three years, to achieve a three-year rolling average outperformance of 1%.
4. After year 1, in any year, to have less than 40% probability of underperforming the benchmark.

The total expected relative return and the total tracking error together enable the calculation of the probability distribution of fund relative returns and the probability of achieving particular values. These are shown, respectively, in Figure 9.6 and Table 9.11. When compared with the structure only distribution, the total distribution is centred further to the left (the expected return is lower) and is much flatter (riskier). The total area under the two curves remains constant at one (the total probability). The flattening has the effect of making the objective of above average performance *easier* to achieve (see Table 9.11). However, it also substantially increases the probability (the area under the curve) of very low relative returns: with structure only, the probability of a relative return of less than −4.5% is 0.01; with structure and stock it is 0.38. Without remedial action, the

Figure 9.6► Distribution of MESS Fund relative returns (structure alone and total)

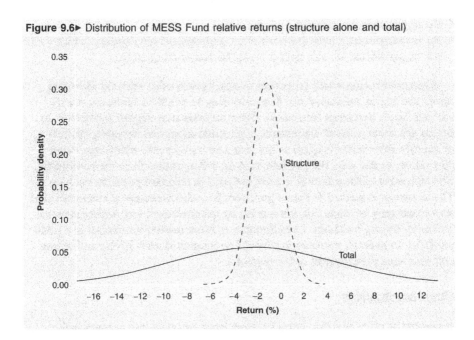

Table 9.11► Probability of achieving at least a given relative return

Relative return (at least)	Probability	
	Structure	Total
-4.50	0.99	0.62
-3.00	0.92	0.53
-1.50	0.59	0.45
0.00	0.17	0.36
1.50	0.02	0.29

fund is unlikely to achieve objective 1 (the probability of doing so is only 0.36), does not achieve objective 4 (the probability of underperforming the benchmark is 0.64), and has little prospect of achieving objectives 2 and 3.

Stock characteristics

While the analyses above provide clear guidance on the restructuring required for a portfolio, a strategy must take into account the stock characteristics. Particular features of the stock may mean that some buildings should not be sold, or at least not immediately, despite the need to change the sub-market weightings. Examples include:

► the need to complete rent reviews which may increase the passing rent above current market levels and so increase the market value[12]
► potential marriage value (from combining two or more property interests) which may be exploited by development

- ▸ the opportunity to refurbish and add value
- ▸ the investment may be in the form of an uncompleted development which could not be considered for sale until it would be nearer completion.

When considering which properties to buy, factors other than the identified sector and region are important. Forecasts may be available at the town level and will direct purchases (and sales). Other analyses may suggest sub-sectors to buy or sell (such as retail warehouses, high street shops and shopping centres) or identify other features (such as size, age and lease length) where mispricing may occur. In this way, the portfolio may be differentiated from the benchmark, although most of these factors are not included in standard portfolio analyses. This is known as a *tilted* or *biased* portfolio. It is also necessary to remember that investment may be made in a sub-market by refurbishment and development as well as by buying buildings. Consideration of these matters is essential in a small portfolio. In general, restructuring should be integrated with buying and selling and with asset management of the portfolio.

Other practicalities

A number of other practical considerations must be taken into account before determining a final strategy.

Lot size
The average lot size varies across sub-markets: it is generally large in capital city office markets and small in provincial industrial markets. For a small portfolio, such as the MESS Fund, with a total value of £100m, this means that exposure to a particular market may be in units of, say £5m or 5%, and that thus the theoretical optimum allocation of, say 2.5%, in a particular sub-market is not possible.

A further problem for small portfolios is that an investment of, say £30m, in one building creates a substantial amount of specific risk. In such circumstances, an upper limit on the value of any one property of, say £10m, is appropriate. On the other hand, investment in a large number of very small properties may prevent access to even medium value markets and will create additional management costs. Accordingly, a minimum lot size of, say £1m, is required. It may also be appropriate to set an average property value of, say £3–4m.

Returns beyond one year
While strategy is set for one year ahead and performance is measured and strategy reviewed annually, forecasts of subsequent years are also required. There is, for example, little point in taking substantial bets, which could not be changed easily, if these were forecast to generate above benchmark returns one year but substantial underperformance in subsequent years.

Timing
The timing of any change in strategy is important. It is necessary to anticipate market movements and to buy and sell at the most advantageous times.

New money

The scope for changing the shape of a fund will depend on whether new money is coming into the fund, or sales are required so that money can be withdrawn. The former creates opportunities; the latter means problems.

Market conditions

If market conditions are poor and liquidity is low, the lead time on a sale may be high and strategies for sales may be difficult to implement. On the other hand, such conditions create a buyers' market in which new money in the fund can result in above benchmark performance.

Market size

It is necessary to consider the size of the market when setting a buy or sell strategy. Substantial buying and selling may move a market against the fund. This is more of a consideration for large funds.

Costs

The cost of sales and purchases should be included in the analysis. Although these have been excluded from the illustrative examples in this chapter, in practice they must be included. In some countries, such as France, high transaction costs create particular problems.

Implementation and monitoring

In arriving at an appropriate strategy, it is essential to ensure that the fund is able to implement it, for example, that it has the staff resources to do so. The simplified analyses set out here assume that all restructuring (and consequent advantages) could be achieved at the beginning of each year, although, in practice, this assumption can be relaxed. The implementation requires to be monitored and adjusted according to changing circumstances.

Strategy for the MESS Fund

The following illustrates the type of strategy which will enable the MESS Fund to achieve its objectives.

Proposed new structure

The proposed new structure and fund bets are shown in Table 9.12 and Figure 9.7. These are derived from an MPT analysis using relative return and tracking error attributable to the structure. The output of this analysis has to be modified to take account of lot size and practical stock issues (see below).[13] The following changes have been made:

1. The negative bets in the three retail markets have been changed to two positives and one neutral.
2. The overall positive bet in the office market has been changed to negative.
3. The positive bet in the Central office market has been substantially reduced.
4. The positive bet in the Southern office market has been changed to negative.
5. The overall positive bet in the industrial market has been reduced.

Table 9.12► Proposed new structure for the MESS Fund

Fund structure

	Retail	Offices	Industrials	Housing	Total
North	10.0%	6.0%	4.0%	0.0%	20.0%
Central	16.0%	11.0%	0.0%	2.0%	29.0%
South	20.0%	20.0%	7.0%	4.0%	51.0%
Total	46.0%	37.0%	11.0%	6.0%	100.00%

Fund bets

	Retail	Offices	Industrials	Housing	Total
North	0.0%	1.0%	2.0%	−3.0%	0.0%
Central	1.0%	1.0%	−2.0%	−1.0%	−1.0%
South	5.0%	−5.0%	1.0%	0.0%	1.0%
Total	6.0%	−3.0%	1.0%	−4.0%	0.0%

Contributions to performance

	Retail	Offices	Industrials	Housing	Total
North	0.00%	0.04%	0.08%	−0.15%	−0.03%
Central	0.08%	0.02%	0.00%	−0.04%	0.06%
South	0.60%	0.45%	0.02%	0.00%	0.87%
Total	0.68%	0.51%	0.10%	−0.19%	0.90%

Figure 9.7► New MESS Fund bets and forecast returns

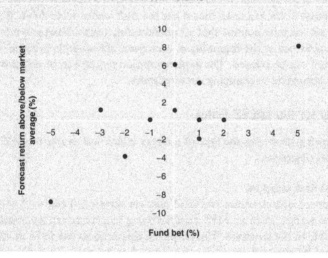

6. The small negative bet in the Northern industrial market has been changed to positive.
7. The neutral position in Central industrials has been changed to negative.
8. The positive bet in the Southern industrial market has been reduced.
9. Although the overall bet in the housing market remains unchanged, the negative bet in the Northern housing market has been increased.
10. The negative bet in the Southern housing market has been removed.

The consequence of these changes is that the contributions of each sub-market bet to relative expected return are now predominantly positive. The structure component of expected relative return has increased from -1.2% to $+0.9\%$. The structure component of risk has decreased from 1.3% to 0.37%.[14]

Buying, selling and asset management
The restructuring has been achieved by the following buying, selling and asset management strategy:

- Northern retail: buy one property for £2m.
- Central retail: buy two properties for £3m each.
- Southern retail: sell underperforming property for £4m and buy three new properties for £4m each.
- Northern offices: no action.
- Central offices: sell underperforming property for £6m and refurbish/redevelop remaining property for £2m.
- Southern offices: sell £30m property which creates too much specific risk and has underperformed and buy two properties for £10m each.
- Northern industrials: sell underperforming property for £1m and buy two replacements for £2m.
- Central industrials: sell underperforming property for £2m.
- Southern industrials: sell one property for £5m and buy one for £2m.
- Northern housing: sell one property for £2m.
- Central housing: no action.
- Southern housing: buy one property for £2m.

This strategy is considered realistic. It is estimated that the combined effect of transactions and asset management will change the stock contribution to relative return from -1.2% to $+0.8\%$ in the first year.[15] Further action and improvements are planned for subsequent years. The number of properties will increase from 25 to 30 and the range of values will decrease from £1–30m to £2–10m (see Table 9.13): both have the effect of reducing specific risk, from 6.9% to 4.6% (see Table 9.14).

Table 9.13► Proposed numbers and sizes of properties in the MESS Fund

Number of properties

	Retail	Offices	Industrials	Housing	Total
North	5	1	2	0	8
Central	8	1	0	1	10
South	6	2	2	2	12
Total	19	4	4	3	30

Average value (£m)

	Retail	Offices	Industrials	Housing	Total
North	2.0	6.0	2.0	0.0	2.5
Central	2.0	11.0	0.0	2.0	2.9
South	3.3	10.0	3.5	2.0	4.2
Total	2.4	9.2	2.8	2.0	3.3

Table 9.14▸ The impact of the new MESS Fund strategy

	Before	After
Structure return	−1.2	0.9
Specific return	−1.2	0.8
Total expected return	−2.4	1.7
Structure TE	1.28	0.37
Specific TE	6.86	4.56
Total TE	6.98	4.58

Figure 9.8▸ Old and new distributions of MESS Fund relative returns

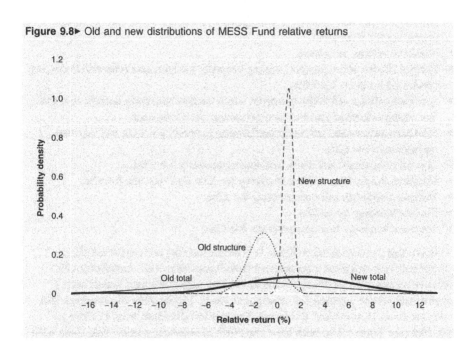

The strategy has a beneficial impact on the structure and stock components of both expected relative return and tracking error, as shown in Figure 9.8. Both the structure alone and the structure plus stock distributions are moved to the right (expected relative return increases) and are narrower (tracking error decreases). Table 9.14 shows that the net effect of the strategy is to increase the expected relative return by 4.1% (from −2.4% to +1.7%) and to reduce tracking error from 7.0% to 4.6%. The probability of achieving the first-year return objective increases from 36% to 64% (Table 9.15). Overall the fund is now in good shape to meet its medium-term objectives. This has been achieved through a combination of structure and stock improvements. It should, however, be re-emphasised that this analysis is illustrative and ignores transaction costs. In practice, these have to be set against expected short-term and long-term increases in return.

Table 9.15► Before and after probabilities of achieving at least a given relative return

Return (at least)	Probability	
	Before	After
−4.50	0.62	0.91
−3.00	0.53	0.85
−1.50	0.45	0.76
0.00	0.36	0.64
1.50	0.29	0.52

9.6 Summary and conclusions

This chapter has applied much of the more theoretical material of earlier chapters to the practical construction and management of a property portfolio. Such active management is in contrast with passive management, where buying and selling is undertaken only to maintain a portfolio's characteristics. The most common form of passive management is an index fund which matches the structure of an index and so is guaranteed to provide index performance, less management costs. Such funds are not possible in the property market as each property is unique and carries with it specific risk, although this may be substantially diversified away in a large portfolio. Passive management is consistent with the Efficient Markets Hypothesis, while active management relies on mispricing in investment markets. Whatever the arguments for or against active management, it dominates investment management.

Fund managers gain more money to manage, and investing institutions gain insurance or pension business if their performance is better than that of their competitors. Thus, fund objectives are set relative to a benchmark of competitors. In this context, return and risk are important not in absolute terms but relative to competitors. Risk is measured by the tracking error, the standard deviation of expected returns relative to the benchmark. Fund management may be considered to be analogous to business planning. It is important to understand:

► what it wants to achieve (objectives)
► where the fund is now (its structure)
► how it got there and what it was good and bad at (performance analysis)
► opportunities and constraints (current structure, stock, market conditions, need to release money by sales and so on)
► how it will achieve the objectives (restructuring, stock selection and asset management).

Objectives should comprise four parts: the benchmark against which to measure fund performance; return relative to the benchmark; risk relative to the benchmark; and timescale for achieving the objectives. Current structure and past performance are considered relative to the benchmark. Performance analysis helps the fund managers understand their strengths and weaknesses in terms of the broad structure of the portfolio (by sector and region), the quality of the stock, and the contributions to returns of transactions, developments and asset management.

Value is added to a property portfolio in three ways:

▶ by taking long/short positions or bets relative to the benchmark in sub-markets expected to perform above/below average
▶ by holding buildings which perform above average for their sub-market
▶ by asset management of the stock.

These all require forecasting ability and, while each may generate positive returns relative to the benchmark, they all create tracking error risk which has to be efficiently managed. Expected relative return and tracking error risk can be forecast and combined to estimate the probability distribution of expected relative returns. This allows the probability of achieving a particular level of relative return to be calculated.

The larger the fund the more important is structure and the less important is stock to return and risk. For such funds, while it is unreasonable to expect to hold a large number of properties with above-average expected return, specific risk can be diversified. However, the structure bets in a large fund are difficult to change because of the large volume of buying and selling required to achieve even a modest restructuring.

The restructuring required to meet fund objectives provides the context for both a buying and selling strategy and for asset management of existing stock. For small portfolios, structure and stock must be considered together and the process of strategy formulations is iterative. Analysis suggests what restructuring is required but this may be constrained by stock considerations such as the need to complete a development or undertake a rent review before sale. Other considerations include:

▶ The need to forecast returns beyond one year so that substantial bets are not in markets forecast to do well in one year but poorly the next.
▶ The need to anticipate market movements and to buy and sell at the most advantageous moments.
▶ The scope for changing the shape of a fund will depend on whether new money is coming into the fund, or sales are required so that money can be withdrawn.
▶ If market conditions are poor and liquidity is low, the lead time on a sale may be high and strategies for sales may be difficult to implement.
▶ If a market is small, substantial buying and selling may move that market against the fund.
▶ When applied in practice, the cost of sales and purchases should be included in the analysis.
▶ It is essential to ensure that the fund is able to implement the strategy.

No matter whether the portfolio is large or small, restructuring, transactions and active management must be undertaken in a co-ordinated and consistent way and the implementation must be monitored and adjusted according to changing circumstances.

The procedures outlined in this chapter have become common in property fund management in recent years. They offer scope to increase returns and to manage risk. Importantly, they apply techniques from other investment markets and enable property to be compared in a consistent way with other investments.

Notes

1. Passive management also involves establishing portfolios with other characteristics such as high- or low-income yields and maintaining these characteristics.
2. If the standard deviation of fund performance were 4%, it would take an average outperformance of 0.5% over 16 years for the fund's performance to be significantly different from the market.
3. The importance of the sectors varies from country to country: in the UK, housing is not a significant institutional investment whereas in countries such as the Netherlands and Switzerland it is. The geographical divisions depend on the size of the country and the availability of data. In the UK, it is standard practice to use the 11 economic planning regions with various subdivisions of London depending on the sector: thus, West Midland offices is one category.
4. Performance measurement services have developed which examine fund manager performance relative to competitors. However, the published figures focus on the returns and not on the risks taken to obtain these returns. The shrewder investor will seek a fund which has had consistently above-average performance rather than one which did very well last year as such performance is unlikely to be repeated.
5. For institutional funds with liabilities, the neutral or benchmark position at the mixed-asset level (shares, bonds and property) would also be driven by actuarial considerations such as duration (see Chapters 3 and 7). However, competitor neutral positions are likely to be driven by the same considerations. This is not a factor at the level of the individual asset portfolio.
6. For mixed-asset portfolios it is common to use economic scenarios to estimate expected risk (see Chapter 3). Scenarios are developed and assigned probabilities. Asset returns are forecast under each scenario and the expected returns and risk are calculated. Perhaps because of the large number of categories, the method is not used for property portfolios and the risk analyses outlined in this chapter are the norm.
7. The structure component can be further divided into sector effects, region effects and combined sector/region effects. The analysis assumes that all the systematic effects are explained by sector and region and that all remaining differences are attributable to stock characteristics. Caution is required when interpreting such attribution analyses. It is possible that other dimensions, such as age and size of the buildings, lease length or town location have systematic effects on returns, but information on these characteristics is less easy to obtain and forecasts of such sub-markets would be very difficult. An analysis of these other components of performance can be used to direct stock selection, for example 'buy shopping centres in Edinburgh'.
8. If the overall impact of transactions is consistently positive for funds in the IPD, there must be other buyers and sellers not in the IPD for whom the impact is equivalently negative.
9. Giliberto et al. (1999) propose correlations which are conditional on the economic environment. They calculate these for mixed-asset portfolios but do not extend the work to property portfolios.
10. It is assumed that the benchmark is well-diversified and contains no specific risk. If this is not true and information on benchmark specific risk is available, it can easily be incorporated into the analysis.

11. For any given property-specific risk and number of buildings, it can be shown that the portfolio-specific risk is at a minimum when the buildings are of equal value.
12. This involves an element of risk as the new rent may be below market expectations. If market expectations are confirmed, the seller will have borne the risk and increased the value of the property. The review may also increase the rent above the market's expectations.
13. For example, buildings will not be available for the exact amounts required by the theoretical allocations in MPT.
14. The minimum theoretical level using an MPT analysis is 0.35%. Lot size and stock issues make this minimum level unachievable.
15. This is a notional figure rather than a calculation.

Further reading

Corgel, J. B., Smith, H. C. and Ling, D. C. (1998) *Real estate perspectives: an introduction to real estate*, Irwin/McGraw-Hill, Boston (third edition): Chapter 8.

Del Casino, J. J. (1995) Portfolio diversification considerations, in *The handbook of real estate portfolio management*, Pagliari, J. L. (Ed.), Irwin, Burr Ridge (IL), 912–66: Chapter 23.

Dubben, N. and Sayce, S. (1991) *Property portfolio management: an introduction*, Routledge, London: Chapter 9.

Hoesli, M. Lizieri. C. M. and MacGregor, B. D. (1997) The spatial dimensions of the investment performance of UK commercial property, *Urban Studies*, 34(9), 1475–94.

Hudson-Wilson, S. (1994) Real estate portfolio management – SampCo: a hypothetical portfolio analysis, in *Managing real estate portfolios*, Hudson-Wilson, S. and Wurtzebach, C. H. (Eds), Irwin, Burr Ridge (IL), 207–36: Chapter 6, Part III.

Lieblich, F. (1995) The real estate portfolio management process, in *The handbook of real estate portfolio management*, Pagliari, J. L. (Ed.), Irwin, Burr Ridge (IL), 998–1058: Chapter 25.

Lightner, C. R. (1994) Real estate portfolio management – What saved PRISA? A case study, in *Managing real estate portfolios*, Hudson-Wilson, S. and Wurtzebach, C. H. (Eds), Irwin, Burr Ridge (IL), 185–207: Chapter 6, Part II.

Mueller, G. R. and Louargand, M. A. (1995) Developing a portfolio strategy, in *The handbook of real estate portfolio management*, Pagliari, J. L. (Ed.), Irwin, Burr Ridge (IL), 967–97: Chapter 24.

Wurtzebach, C. H. (1994) Real estate portfolio management – Financial theory replaces anecdote, in *Managing real estate portfolios*, Hudson-Wilson, S. and Wurtzebach, C. H. (Eds), Irwin, Burr Ridge (IL), 165–84: Chapter 6, Part I.

Part III
Property in a wider context

Chapter 10

Property in mixed-asset portfolios

10.1 Introduction

In Chapter 4, methods of constructing property indices were presented. In Chapter 7, Modern Portfolio Theory (MPT) was presented. The optimal diversification of property portfolios was discussed in Chapter 9. This chapter considers the role of property in multi- or mixed-asset portfolios – that is, those containing shares, bonds and property. Among other things, the optimal weight which should be allocated to property is analysed. This step, which is known as inter-asset diversification, is very important and has a much greater impact on the overall portfolio performance than the decision as to how to invest within each asset category (intra-asset diversification).

The benefits from including property in mixed-asset portfolios are examined in a Modern Portfolio Theory (MPT) framework. Traditionally, an efficient frontier of portfolios containing shares and bonds only is constructed first. Then, a frontier which contains combinations of shares, bonds and property is constructed. If the frontier containing shares, bonds and property lies well above the frontier containing only shares and bonds, then property has a positive role to play in diversifying mixed-asset portfolios. This means that when property is included, for a given return level, the risk is lower, or, equivalently, that for a given risk level, the return is higher.

In order to construct such frontiers, data are needed on the expected return and risk of the various asset classes. Correlation coefficients between each pair of assets are also needed. These figures can be estimated from return indices for the various asset classes. As discussed in Chapter 4, such indices are readily available for shares and bonds. In the case of property, indices are more difficult to construct due to the characteristics of this asset class. Several different types of index exist, the natures of which vary substantially.

In some countries, such as the UK, institutions predominantly consider commercial property (mainly shops, offices and industrials) as the means of investing in property. In most other countries, residential properties are also considered by institutions. In fact, apartment buildings constitute as much as 46% of the property holdings of French insurance companies and 74% of those of Swiss pension funds (Dumortier, 1997; Schärer, 1997). Institutions can also consider shares in property companies as a way of gaining exposure to the property market (see Chapter 11). Each of these three types of exposure to the property market (commercial, residential and property company shares) is examined separately in this chapter.

For commercial property, little information exists on transaction prices, and appraisal-based indices are usually constructed for this type of property. As discussed in Chapter 4, such indices suffer from smoothing and the variances of property and the covariances of property with other assets are biased towards zero. Such a phenomenon has to be controlled for when the impact of including commercial property in multi-asset portfolios is examined. For residential properties, appraisal-based indices are used (in the USA and the Netherlands, for example), but transaction data are more readily available and indices of transaction prices can be used. For indirect property, several indices exist (see Chapter 4).

This chapter is organised as follows. First, the role of commercial property in multi-asset portfolios is examined in Section 10.2. Then, Sections 10.3 and 10.4 consider, respectively, the role of residential properties and property company shares in diversifying mixed-asset portfolios. In Section 10.5, the results from the various studies presented in Sections 10.2 to 10.4 are compared to the actual holdings of institutions, and some possible reasons for the observed differences are discussed. These reasons include limitations of data, illiquidity, transaction costs, size of the fund, omitted assets, competitor risk and liabilities. Finally, Section 10.6 contains a summary and conclusions.

10.2 Commercial property in the portfolio

As discussed in Chapters 1 and 2, the term 'commercial property' in the UK includes shops, offices and industrials. In the USA, it also includes residential properties (residential buildings and hotels) held for investment purposes. In the UK, only appraisal-based indices are available for commercial property. In the USA, commercial property (including residential buildings) is mainly monitored through an appraisal-based index (the NCREIF Property Index, NPI). As of the second quarter of 1998, the NPI comprised 84% of commercial properties (in the UK sense) and 16% of residential properties.

In the USA, some attempts have been made to construct and use indices of transaction prices for commercial property. This section presents results based on appraisal-based indices (mainly for the USA and the UK, but also for Australia, Canada, New Zealand and South Africa) and results based on transaction data (for the USA).

The expected return and risk of each asset class and the correlation coefficients between each pair of assets are needed in order to construct efficient frontiers (see Chapter 7). The return and risk characteristics of commercial property and a comparison of these parameters with those of shares and bonds are presented first. Then, the correlation coefficients are discussed. Finally, the results of the efficient frontiers analysis are considered. A similar structure is used in Sections 10.3 and 10.4.

The values of the historical estimates of average return, risk and correlation parameters will depend on the time period considered. The Modern Portfolio Theory (MPT) results will be sensitive to the value given to these parameters. It is, therefore, of interest to consider various time periods, but also different countries to test the robustness of the results.

Return and risk with appraisal-based indices

The analysis of the return and risk of property has received much attention in the literature (for a review, see Norman et al., 1995). When appraisal-based real estate series are used, the standard deviation of property returns is very low (see also Chapter 4). Ibbotson and Siegel (1984), for example, report the following figures for the USA for the period 1947–82: the average annual return and standard deviation are 8.3% and 3.7% for property (comprising farms, residential properties and business real estate); 12.4% and 17.5% for shares; and 3.4% and 9.7% for bonds.[1]

More recently, Coleman et al. (1994) report an average quarterly return and a standard deviation of 2.83% and 2.47% for property, 4.22% and 5.73% for stocks and 2.70% and 2.09% for bonds, respectively, for the period from the first quarter of 1978 to the fourth quarter of 1989. The property returns are measured from the Russell–NCREIF index (now NCREIF Property Index, NPI) (see Chapter 4).

These results suggest that property constitutes a very interesting asset class. In the study by Ibbotson and Siegel (1984), property's average return is higher than that of shares, whereas its standard deviation is lower. In the study by Coleman et al. (1994), the coefficient of variation (which measures the relative dispersion of the returns of the various investment classes and which is measured as standard deviation/mean return) amounts to 0.87 for property, 1.36 for shares and 0.77 for bonds, respectively, indicating more attractive return-risk characteristics for property than for shares. As mentioned above, the results for the USA mainly concern commercial property, but also some residential property.

For the UK, MacGregor and Nanthakumaran (1992) report quarterly average returns and standard deviations of real returns for commercial property, shares and gilts (UK government bonds) for the period 1977–90. The property returns are computed on the basis of the JLW index. The average return and standard deviation are 1.63% and 2.45% for property, 2.80% and 9.90% for shares and 1.03% and 6.94% for gilts, respectively. The coefficients of variation are 1.50, 3.54 and 6.74 for property, shares and gilts, respectively.

These results indicate that commercial property in the UK also exhibits attractive return-risk characteristics when appraisal-based series are used. From a mean-variance criterion perspective, the results of MacGregor and Nanthakumaran (1992) show that property dominates bonds in the UK, a result which is in line with that reported by Ibbotson and Siegel (1984).[2]

Studies have also been undertaken for other countries, such as Australia, Canada, New Zealand and South Africa. Among other countries, Newell and Webb (1996) report semi-annual return and risk figures for Australia and Canada

Table 10.1► Average return and risk for property, shares and bonds in Australia and Canada, 1985–93

	Average return		Risk (SD)	
	Australia	Canada	Australia	Canada
Property	5.32%	3.41%	6.99%	4.91%
Shares	9.57%	4.92%	14.51%	9.84%
Bonds	9.05%	6.80%	6.71%	5.13%

Source: Newell and Webb (1996).
Note: The returns are semi-annual.

Table 10.2► Average return and risk for property, shares and bonds in New Zealand and South Africa[3]

	Average return		Risk (SD)	
	New Zealand (1990–95)	South Africa (1980–95)	New Zealand (1990–95)	South Africa (1980–95)
Property	–0.70%	13.0%	5.72%	7.83%
Shares	3.99%	16.2%	16.17%	21.23%
Bonds	5.63%	14.8%	3.90%	1.93%

Source: Newell and Webb (1998).
Note: The returns are semi-annual for New Zealand and annual for South Africa.

for the period 1985–93 (see Table 10.1). Newell and Webb (1998) report semi-annual return and risk figures for New Zealand offices for the period 1990–5, and annual figures for South African commercial property for the period 1980–95 (see Table 10.2).

The results for these four countries are quite different from those observed in the USA and the UK. From a mean-variance perspective criterion, property is dominated by bonds in Australia, New Zealand and South Africa. In Canada, property appears to provide a lower return and risk than bonds. In all four countries, property exhibits lower return and risk parameters than stocks. Caution should, however, be exercised when interpreting these results as the data cover only short time periods which do not span over a full property cycle (1985–93 for Australia and Canada, and 1990–5 for New Zealand). When the US and UK results are considered for the period 1985–93, property exhibits lower return and risk attributes than both shares and bonds. Further, the South African results are only for capital returns and not total returns.

As discussed in Chapter 4, appraisal-based indices have been shown to be smoothed, which leads the variance to be biased towards zero. Accordingly, several authors have unsmoothed appraisal-based series in order to obtain more reliable variances (and hence standard deviations). Ross and Zisler (1991), for instance, suggest that the standard deviation of property should be increased by a factor of 3 to 5 for US data using quarterly nominal returns on the Russell–NCREIF index.

Fisher et al. (1994) unsmooth the Russell–NCREIF index under two assumptions: (1) the underlying true returns are uncorrelated across time,

Table 10.3► Factors reported in the literature to convert standard deviation from smoothed series to 'true' standard deviation[4]

Study	Country	Index	Nominal or real	Returns frequency	Time period	Factor
(1)	USA	Russell–NCREIF	Nominal	Quarterly	1978–85	3 to 5
(2)	USA	Russell–NCREIF	Real	Annual	1979–92	1.58–1.66
(3)	UK	IPD	Nominal	Monthly	1987–90	3.44
(4)	UK	JLW	Nominal	Quarterly	1977–90	2.86
			Real	Quarterly	1977–90	1.92
(5)	UK	JLW	Nominal	Annual	1970–92	1.57
(6)	UK	IPD	Nominal	Annual	1987–92	3.45

Sources: (1) Ross and Zisler (1991); (2) Fisher *et al.* (1994); (3) Brown (1991); (4) MacGregor and Nanthakumaran (1992); (5) Barkham and Geltner (1994); (6) Newell and MacFarlane (1996).

consistent with the classical hypothesis of weak-form informational efficiency in asset markets, and (2) property markets are not informationally efficient in the same way as securities markets, therefore, the assumption of unpredictability may not be valid for true property returns. The standard deviation of real annual capital returns on property rises from 5.20% to 8.62% under the first assumption and to 8.19% under the second assumption.

MacGregor and Nanthakumaran (1992) unsmooth the JLW property series. The standard deviation of real returns rises from 2.45% to 4.70% with the unsmoothed series and the coefficient of variation for property rises from 1.50 to 3.20, compared to a coefficient of 3.54 for shares and of 6.74 for gilts. A similar analysis of the nominal quarterly series suggests an increase in the standard deviation by a factor of nearly 3 when the unsmoothed series is used. Also using the JLW index, Barkham and Geltner (1994) find that the annual standard deviation of property for the period 1970–92 rises from 11.8% to 18.6% (an increase by a factor of 1.57) when the series is unsmoothed using the middle value for the estimated quarterly valuation smoothing parameter.

Brown's (1991) analysis of IPD monthly nominal returns shows a factor of 3.44. Newell and MacFarlane (1996) also use the IPD monthly return series for the 1987–92 period, and report that the unadjusted volatility estimates need to be increased by a factor of approximately 3.5. The multipliers reported in the literature are summarised in Table 10.3. As can be seen, the multipliers vary across countries and depend on the time period considered. As pointed out by MacGregor and Nanthakumaran (1992), other data series are almost certain to have different multipliers.[5]

Even with unsmoothed appraisal-based series, property appears to be an attractive asset class. Corgel *et al.* (1998), for instance, report an annual average return and standard deviation of 12.8% and 7.6%, respectively, for US property over the period from 1978 to the second quarter of 1995. The figures are 16.2% and 14.8% for shares and 10.4% and 8.4% for bonds. These results suggest that property dominates bonds from a mean-variance criterion perspective.[6] MacGregor and Nanthakumaran (1992) also find that property dominates bonds even when smoothing has been removed from the index. In all cases, property has a lower return and risk than shares.

The return and risk attributes of property thus remain attractive once the series have been unsmoothed. Two reasons can be given to explain the fact that property dominates bonds from a mean-variance criterion perspective. First, and as discussed in Chapter 2, property investments are illiquid investments, and an illiquidity premium is incorporated in expected returns which needs to be subtracted from *ex post* returns. This would make the comparison with other asset classes more meaningful as the varying liquidity levels would be accounted for. MacGregor and Nanthakumaran (1992), for instance, use a 50 basis point quarterly illiquidity premium which they subtract from the average return on property. With the unsmoothed series and the illiquidity premium, property appears to provide a lower return and risk than both shares and bonds.

Another reason which can explain the observed dominance of property over bonds in a mean-variance context is that the property indices contain ungeared (unleveraged) properties, whereas many investors would consider debt financing when purchasing property. If geared properties are considered, both return and risk will increase.

Ahern *et al.* (1998) argue that leveraged real estate is a 'short position in a mortgage combined with unleveraged real estate'. 'The use of leverage can be thought of as countering the bond-like characteristics of real estate, or simply, as "debonding" real estate.' For example, assuming a 10.5% return on an unleveraged real estate investment and a 7.5% mortgage interest rate, the expected return would be 13.5% with a loan-to-value ratio of 50%, and 22.5% with a loan-to-value ratio of 80%. Also, a 50% loan-to-value ratio would approximately double the standard deviation.

The overall conclusion of these studies is that property appears to have a higher return and risk than bonds and a lower return and risk than shares. As will be seen below, this general conclusion also holds when transaction-based indices of property are considered.

Return and risk with transaction-based indices

Some US studies have made use of transaction data in order to construct indices of commercial property. However, very few studies use averages of transaction prices unadjusted for the quality of the properties in order to measure the return and risk of property because these series may not reflect the true evolution of prices from one period to another. As mentioned in Chapter 4, the quality of the buildings included in the sample one year may be dramatically different from the quality of the buildings included in the sample another year. This can lead to serious biases: the index can rise when property prices have declined, or the index can drop when property prices have risen.

An unadjusted price index of commercial property is constructed, for example, by Fisher *et al.* (1994), mainly for comparison purposes with other property price indices. Such an index is constructed by dividing the Russell–NCREIF historical Net Operating Income (NOI) series by the transaction-based capitalisation rates reported by the American Council of Life Insurers (ACLI). Fisher *et al.* (1994) find an average return of 0.10% and a standard deviation of 9.36% for the period 1979–91. For US shares, they find an average of 10.72% and a standard deviation of 12.39% on the basis of the returns on the S&P500 for the period 1979–92 (only the price change component of returns is considered). This would suggest that US property exhibits lower return and risk characteristics than shares.

Hedonic indices of commercial property are constructed for the USA by Cole (1988) and by Webb et al. (1992).[7] Cole (1988) reports for the period 1982–6 an average return of 11.4% and a standard deviation of 10.2% for property. With an appraisal-based series, the average return is 11.1% and the standard deviation 2.1%. This result suggests that appraisal-based series yield a good estimate of the average return on property. For shares, the average return is 21.1% and the standard deviation is 15.4%. According to this study, property seems to yield a lower return and risk than an investment in shares. The study by Webb et al. (1992) is somewhat puzzling in that they report, for the period 1980–8, a lower standard deviation for the hedonic property series than for the appraisal-based series.

Fisher et al. (1994) construct a hedonic index for the period 1982–92 and report an average real return of −0.38% and a standard deviation of 14.42%. For shares, these figures are 10.72% and 12.39%, respectively. According to this study, property yields a lower return and a higher risk than shares.

As reported by Fisher et al. (1994), even with a perfect hedonic model, smoothing due to temporal aggregation still occurs in the index, as properties that sell at different times during the year are being used to estimate values as of a single point in time. In addition, estimation error injects random noise into the hedonic index, artificially adding non-systematic volatility into the returns. The greater short-run volatility (14.42%) and negative first-order autocorrelation (−9.4%) in the annual real returns of the hedonic index suggests that the random estimation error noise is dominating the temporal aggregation smoothing. This reflects the relative paucity of transaction observations in each year.

Gatzlaff and Geltner (1998) construct a repeat sales index for commercial property in Florida for the 1982–96 period. Their data set includes 4373 properties. The repeat sales index is compared to the NCREIF Property Index (NPI) for Florida. Surprisingly, the repeat sales index only shows a little more volatility than the appraisal-based NPI (standard deviation of 4.07% for the former and 3.86% for the latter).

Although more research is needed with transaction-based data, the general conclusion is that property exhibits a higher return and risk than bonds and a lower return and risk than shares.

Correlation with shares and bonds

Low positive or negative correlations between property returns and the returns on shares and bonds are observed both in the USA and the UK. Hartzell et al. (1986) report a correlation coefficient of −0.12 between property returns and share returns, and of −0.39 between property returns and bond returns for the USA over the period 1973–83. Over the period 1979–92, Fisher et al. (1994)

report a correlation coefficient of –0.04 between returns computed from the Russell–NCREIF index and the returns on the S&P500 index. For the UK, MacGregor and Nanthakumaran (1992) report a correlation coefficient of –0.19 between annual real returns on property and shares and a correlation of 0.01 between property and gilts for the period 1977–90.

For the period 1968–90, however, MacGregor and Nanthakumaran (1992) report much higher (and positive) correlation coefficients: 0.28 between property and shares, and 0.29 between property and gilts. Thus, correlation coefficients are not stable over time, and it is clear that single years have a dramatic effect on the correlation. In the case of the reported figures for the UK, the inclusion of the slump years of the mid-1970s reduces the property/share correlation from 0.28 to –0.19.

Some authors have argued that, owing to the smoothing of appraisal-based series, the covariance and correlation coefficient of short-interval returns with contemporaneous returns on other assets is biased towards zero (Geltner, 1993a; Fisher *et al.*, 1994). If unsmoothed property returns are used for the USA over the 1979–92 period, Fisher *et al.* (1994) show that the correlation between property returns and stock returns is –0.11 or –0.07 depending on what hypothesis is made regarding the efficiency of the property market (–0.04 when smoothed returns are used). As can be seen, the results are virtually the same as when the smoothed index is used.

Ziobrowski and Ziobrowski (1997) also report slightly higher magnitudes of correlation coefficients when unsmoothed property returns are used for the period 1970–95. The correlation between property and stocks is between –0.14 and –0.10 depending on the adjustment parameter (–0.07 with smoothed returns) and the correlation between property and Government bonds is between –0.31 and –0.28 (–0.21 with smoothed returns).

For the UK, MacGregor and Nanthakumaran (1992) unsmooth the JLW series and compute correlation coefficients between the unsmoothed property series and stocks and bonds. The quarterly correlation of real returns between property and shares is –0.11, and between property and gilts –0.11. These correlations are exactly the same as when the smoothed JLW series is used.[8] Barkham and Geltner (1994), however, report an increased correlation coefficient between property and stocks when the JLW index is unsmoothed. With a middle value for the quarterly valuation smoothing parameter, the correlation between property and shares rises from 0.22 to 0.38 for the period 1970–92.

Thus, when appraisal-based series are used, the returns on commercial property appear to be lowly correlated with those on shares and bonds. This conclusion holds even when unsmoothed real estate returns are used. Although from a theoretical perspective smoothing should lead correlation coefficients to be biased towards zero, the empirical evidence is somewhat mixed as regards the effect of unsmoothing returns on the magnitude of the correlation coefficient.

Transaction-based indices of commercial property have also been used in order to ascertain the degree of correlation between the returns on commercial property and those on shares and bonds in the USA. Fisher *et al.* (1994) compute a quality-unadjusted property price index and report a correlation of 0.32 between the property series and the S&P500 index of US shares. This result would suggest that property returns are positively but moderately correlated with the returns on shares. For the period 1983–92, however, Fisher *et al.* (1994) report a much higher correlation of 0.50 between the returns computed from their hedonic index and the returns on the S&P500 stock index.[9]

Transaction-based indices generally suggest positive but moderate degrees of correlation between commercial property returns and the returns on shares and bonds, suggesting diversification benefits from including property in mixed-asset portfolios. The observed low correlation coefficients could be due to some temporal aggregation remaining in the series (both in the appraisal-based and hedonic indices) or to supply cycle effects. As property supply cannot adjust quickly to changes in demand, new supply may still be arriving on the market when it is declining. The consequence is that returns are further depressed and the impact may last for years. Further, new construction may not take place in rising markets if there is existing over-supply, with the result that when a supply constraint is reached, returns rise substantially (see Chapter 2 and Ball *et al.*, 1998).

Efficient sets

As discussed in Chapter 7, two things are important for inclusion of a new asset in a portfolio: the correlation between the asset and those already included in the portfolio and the return/risk ratios of the assets. The magnitude and sign of the correlation coefficients have a great impact on the role that can be played by any particular asset in a portfolio. The closer to −1 is the correlation coefficient between an asset and the assets in the portfolio, the more beneficial it will be to include it in a portfolio. On the other hand, if the correlation coefficients are close to 1, then the impact on diversification of including that asset would be marginal. If there is a perfect correlation (that is, if the correlations are equal to +1), the inclusion of that asset would have no effect on diversification.

The return and risk characteristics of each asset class are also important in determining the optimal portfolio compositions. For instance, if an asset with the highest return and risk is added, it will necessarily lie on the efficient frontier no matter what its correlation with other assets is. Bonds usually lie at the bottom end of the frontier and shares at the top end of the frontier, whereas portfolios which contain property lie in the middle range of the frontier.

Commercial property has been reported to have very low standard deviations of returns when appraisal-based series are used. Moreover, property's returns appear to be weakly correlated with the returns on shares and bonds. When such characteristics are used for commercial property in order to ascertain the role that can be played by that asset class in diversifying a portfolio containing shares and bonds, the conclusion can only be that commercial property is a significant contributor in improving a portfolio's performance.[10] The weight which should be allocated to property is now considered.

For the USA, Coleman *et al.* (1994) report an optimal weight of 84.6% for property when the quarterly return target is 3.0%, and of 15.7% when the target is 4.0% for the period from 1978 to 1989. For the UK, MacGregor and Nanthakumaran (1992) compute the optimal weight which should have been allocated to commercial property for the period 1977–90. The weight which should have been attributed to property is 88% for a low return portfolio and is 11% for a high return portfolio.[11] For the period 1968–90, the weight which should have been given to property is 69% and 18%, respectively, depending on the level of return considered. When three-year averages are used for the period 1973–90, the weights are 66% and 12%, respectively, and when five-year averages are used, the optimal weights are 74% and 12%, respectively. These weights, thus, seem to be quite stable for different holding periods.

As mentioned before, the smoothing problem of appraisal-based series leads to the standard deviations of property being underestimated. If this is the case, the weight given to property in portfolio analyses would be overestimated. Several authors have attempted to modify the standard deviation of property in order to gain more accurate knowledge of the weight which should be attributed to property.

One such attempt is made by Webb and Rubens (1987) who multiply the standard deviation as calculated from an appraisal-based series by 3 and by 6, respectively.[12] With a factor of 3, property's weight is in the range 60% to 86%, and with a factor of 6, the weights lie between 43% and 74%.[13] Firstenberg et al. (1988) assume that the standard deviation of property is equal to that of shares. For the period 1978–85, the weight which should have been allocated to property is between 38% and 100%, depending on the level of risk aversion.

MacGregor and Nanthakumaran (1992) unsmooth the JLW series for the period 1977–90. They find that property should account for 67% at the low return and risk end of the frontier and for 12% at the high return and risk end. Barkham and Geltner (1994) find that the optimal mix changes from 58% for property and 42% for stocks to 20% for property and 80% for stocks when the JLW index for the period 1970–92 is unsmoothed (bonds are not considered as being an investment alternative).

However, in these analyses for the UK, as is the case for the studies that modify US property's standard deviation, the illiquidity of property still has not been accounted for. MacGregor and Nanthakumaran (1992) subtract a 50 basis points illiquidity premium from the average return and find that the optimal weight which should be allocated to property does not vary substantially. For a low return portfolio, property should comprise 66% of the portfolio and, for a high return portfolio, this weight should be 11%.

Several US authors (for example, Fogler, 1984; and Ennis and Burik, 1991) have suggested an optimum weight of 15–20% for commercial property in mixed-asset portfolios.[14] This range is confirmed in a recent study by Ziobrowski and Ziobrowski (1997). For the UK, the theoretical weights for commercial property are slightly higher (see MacGregor and Nanthakumaran, 1992; and Brown and Schuck, 1996).

Lee et al. (1996) reverse the optimisation process and analyse the level of return that an asset needs in order to provide a particular portfolio weight. Survey data are used by these authors, who compute the return required from property for it to achieve a holding of 15% in UK pension fund portfolios. Property's required return is less than the expected return, and these authors conclude that a holding of 15% is, therefore, justified.

Table 10.4▶ Average annual return and risk for apartments, commercial property, shares and bonds in the USA (1Q82–4Q93)

	Average return	Risk (SD)
Apartments	2.002%	1.711%
Commercial property	1.259%	1.740%
Shares	4.101%	7.828%
Bonds	3.747%	5.929%

Source: Liang *et al.* (1996).

10.3 Residential property in the portfolio

Return and risk

Institutions in some countries (such as France and Switzerland) invest extensively in the residential property sector. In the USA, the percentage of residential properties in the property holdings of pension funds amounts to 16% (Dumortier, 1997). Because residential properties are viewed as being commercial property in the USA, overall appraisal-based indices in the USA encompass residential property. The weight of residential property in the NCREIF Property Index, NPI (16%) is in line with investment practice by US institutions. In the UK, institutions do not consider residential property as an investment alternative; nonetheless, some empirical studies for the UK have been undertaken.

Some US studies have made use of appraisal-based indices for residential property and have generally reached similar conclusions to those when the commercial sector as a whole is considered. Goetzmann and Ibbotson (1990), for instance, report an average annual rate of return of 8.6% and a standard deviation of 3.0% for the period 1970–86. These figures compare with 10.5% and 18.2%, respectively, for shares and 8.4% and 13.2%, respectively, for long-term bonds. For commercial property, the average return is 10.9% and the standard deviation 2.6%.[15]

Liang *et al.* (1996) calculate the return and risk on apartment investments for the period from the first quarter of 1982 to the fourth quarter of 1993. They use an appraisal-based total return index for apartments. The return and risk parameters of apartment investments are very similar to those of commercial property and, over that period, apartment investments appear to provide a lower return and risk than investments in both shares and bonds (see Table 10.4). The correlation between commercial property returns and the returns on apartments is 0.743.[16] This figure suggests that the two types of property behave in a similar fashion, which would explain why similar results are observed between the two types of property and shares and bonds.

Hutchison (1994) provides an analysis of housing as an investment in the UK. The data on housing represent the opinions of the district valuers derived from sales information in their possession. Table 10.5 contains the return and risk parameters for housing, shares and gilts in the UK from 1984 to 1992. Housing in the UK appears to provide a lower return and risk than an investment in shares and a higher return and risk than an investment in gilts. The figures contained in

Table 10.5▶ Average annual return and risk for housing, shares and gilts in the UK (1984–92)

	Average return	Risk (SD)
Housing	12.09%	12.18%
Shares	17.47%	12.23%
Gilts	11.37%	5.23%

Source: Hutchison (1994).

Table 10.5 suggest that housing did not represent a very good investment alternative as its return is only marginally higher than that on gilts, with risk levels similar to those of shares. When mortgage interest relief is considered, however, the average annual return on housing is 15.10% and the standard deviation of returns 11.61%, making it an attractive investment.

In Switzerland, institutions invest heavily in the residential property sector. The role of residential property in mixed-asset portfolios is, thus, the focus of much research. A hedonic index for apartment buildings in Geneva is constructed by Bender *et al.* (1994) and Hoesli and Hamelink (1996). The average return and standard deviation of property over the 1978–92 period is 7.19% and 13.90%, respectively. For shares, the average return is 8.10% and the standard deviation is 18.85%. For bonds, the figures are 4.33% and 4.18%, respectively. Apartment buildings in Geneva, thus, appear to yield a lower return and risk than shares and a higher return and risk than bonds.

An index of condominiums in Geneva is constructed by Hoesli, Giaccotto and Farvager (1997) for the period 1968–93 using the repeat sales method (see Chapter 4). Condominiums in Geneva yielded a mean return of 4.20% and a standard deviation of 10.08% over that period. For shares, the figures are respectively 7.20% and 21.53%, and for bonds 5.46% and 5.13%. The shares and bonds returns are total returns (that is, returns which include both the income and capital components), whereas the condominium series only measures changes in capital value. If an appropriate adjustment is made, condominiums in Geneva have a lower return and risk than shares and a higher return and risk than bonds.

Thus, in general, residential property has a lower return and risk than shares and a higher return and risk than bonds. This conclusion is consistent with the results obtained for commercial property. The correlation between residential property and shares and bonds is considered next.

Correlation with shares and bonds

Liang *et al.* (1996) calculate correlation coefficients between apartment investments in the USA and investments in shares and bonds. These figures are −0.100 and −0.067, respectively. Hutchison (1994) computes equivalent figures for housing in the UK. The correlation coefficient between housing and shares is 0.0772 and that between housing and gilts is −0.2627.

For Switzerland, Hoesli and Hamelink (1997) compute correlation coefficients between Geneva apartment buildings, Zurich apartment buildings, Swiss stocks and Swiss bonds for the period 1981–92 (see Table 10.6). If the index of

Table 10.6▶ Correlation coefficients between Geneva apartment buildings, Zurich apartment buildings, Swiss shares and Swiss bonds (1981–92)

	Geneva apartment buildings	Zurich apartment buildings	Swiss shares	Swiss bonds
Geneva apartment buildings	1.00			
Zurich apartment buildings	0.41	1.00		
Swiss shares	−0.11	0.18	1.00	
Swiss bonds	−0.33	−0.29	0.15	1.00

Source: Hoesli and Hamelink (1997).

condominiums in Geneva constructed by Hoesli, Giaccotto and Farvager (1997) is used for the period 1969–93, the correlations of apartment buildings with shares and with bonds is effectively zero.

These results suggest that the returns on apartment buildings are lowly correlated with the returns on shares and bonds, a result which is consistent with that reported for commercial property. The correlation coefficient between property (commercial or residential) and shares is usually slightly positive, while the correlation coefficient between property and bonds is most often slightly negative.

Efficient sets

Hoesli and Hamelink (1996) use a hedonic index of apartment buildings in Geneva to analyse the optimal role of such investments in Swiss multi-asset portfolios for the period 1978–92. The illiquidity of property is accounted for by subtracting an illiquidity premium of 50 basis points from the average return on property. They report weights for property between 10% and 57% depending on the level of risk aversion, although for most investors this range should be 20–30%.

Liang *et al.* (1996) construct efficient portfolios for the USA containing shares, bonds, commercial property and apartments. As the return and risk attributes of apartments were slightly more attractive than those of commercial property during the period 1Q1982 to 4Q1993, the entire allocation to property in their multi-asset portfolios is apartments. The optimal weight is in the 15–25% range. This result is important, as institutions in countries where the focus has been exclusively on commercial property could (and probably should) consider apartments as an investment alternative.

10.4 Securitised property in the portfolio

As a result of the relative paucity of transaction-based data in direct property markets, it may seem appealing to proxy the returns and risks of direct investments with data for indirect property investments (these investments are covered in Chapter 11). Indirect property, however, has been shown to be an imperfect indicator of direct markets and to capture some stock market factors. Thus, the results based on indirect property are likely to underestimate the role of

Table 10.7▶ Average monthly return and risk for equity REITs, shares, small-value shares and bonds in the USA (January 1976–June 1993)

	Average monthly return	Risk (SD)
Equity REITs	1.46%	0.0385%
Shares	1.31%	0.0444%
Small-value shares	1.75%	0.0634%
Bonds	0.90%	0.0191%

Source: Mueller *et al.* (1994).

direct property in a mixed-asset portfolio. However, for investors wishing to invest in property through the shares market, the results presented below should prove useful. Such an investment strategy could be the preferred strategy for some investors, for example, because the liquidity of property shares is greater or because access to foreign property markets is easier. It might even be that both direct and indirect investments should be considered. Some preliminary evidence on this issue is also provided below.

It is important to understand the nature of the indirect vehicles before the investment is made. Features such as the tax status and the level of gearing of the funds have to be considered. In particular, the composition of the funds' portfolios (commercial property, residential or a combination of both) has to be analysed. This varies quite substantially from one country to another and often even within countries. Property companies in Switzerland, for instance, invest predominantly in apartment buildings, while UK property companies mostly focus on commercial properties (see Chapter 11).

Return and risk

Mueller *et al.* (1994) compute the average monthly return and standard deviation of equity REITs, stocks, small-value stocks and bonds in the USA for the period from January 1976 to June 1993.[17] These figures are reported in Table 10.7 and clearly show that US equity Real Estate Investment Trusts (REITs) have very similar return and risk characteristics to shares and small-value stocks.[18] Bonds have a lower return and risk. Similar results are reported by Han and Liang (1995) and Brueggeman and Fisher (1997).

Mueller and Laposa (1996) show that the return and risk characteristics of REITs vary quite substantially depending on the property type (apartment, retail, hotel, self-storage, office, industrial, healthcare and manufactured homes). Liang *et al.* (1996) analyse apartment REITs and report an average quarterly return of 4.183% and a standard deviation of 11.955% for apartment REITs for the period from the first quarter of 1982 to the fourth quarter of 1993. For the same period, these figures amount to 3.704% and 7.258%, respectively, for equity REITs.

It has been argued that one of the reasons which could explain why REITs behave in a similar way to shares is gearing (leverage). Fisher *et al.* (1994) ungear equity REIT returns. Even when this is done, the standard deviation of returns (14.25%) is higher than the standard deviation of the unsmoothed property series (8.62% or 8.19%, depending on the desmoothing parameter used). This result suggests that property company shares capture many stock market features.

Table 10.8▶ Average monthly return and risk for property company shares, shares and bonds in the Netherlands, Sweden, Hong Kong, Japan, Singapore and Canada (January 1985–August 1994)

	Property shares		Shares		Bonds	
	Average return	Risk (SD)	Average return	Risk (SD)	Average return	Risk (SD)
Netherlands	0.18%	3.16%	0.74%	4.77%	0.59%	1.27%
Sweden	0.28%	13.00%	1.30%	7.40%	0.92%	1.38%
Hong Kong	2.40%	8.96%	1.85%	8.83%	NA	NA
Japan	0.98%	8.70%	0.52%	6.58%	0.57%	1.89%
Singapore	1.60%	12.08%	1.11%	7.82%	NA	NA
Canada	−1.29%	8.33%	0.48%	4.25%	0.94%	2.48%

NA = Not available.
Source: Eichholtz (1996).

Ghosh *et al.* (1996), however, report two facts which may indicate that property company shares are becoming more like property and less like stocks. First, REITs appear to be less liquid than comparable-size stocks. Second, institutional investment in REITs is growing.

When French property securities are used, the results are quite similar to those for the USA. For property, Hoesli and Thion (1994) report an average annual return of 8.1% and a standard deviation of 13.4% for the period 1982–92. For shares, the average return is 9.1% and the standard deviation is 15.7%, whereas the figures are 5.8% and 3.5%, respectively, for bonds. For the UK, Eichholtz (1996) also finds quite similar return and risk characteristics between property company shares and stocks. For the period from January 1985 to August 1994, the average monthly return is 0.79% and the standard deviation is 6.72% for property shares. The figures are 0.89% and 5.58%, respectively, for shares.

Eichholtz (1996) also reports the return and risk on property companies, stocks and bonds for several other countries (Table 10.8). These figures suggest that property companies are in several countries riskier than stocks, and that the volatility of bonds is substantially lower than the volatility of both property shares and stocks. Caution should, however, be exercised as these results only concern a relatively short time period.

Real estate mutual funds in Switzerland are interesting in that the influence of property features should be stronger than is the case for property securities in most other countries. According to Swiss law, unit fund holders can ask for redemption of their shares at the net asset value (which is based on appraisals). Stock market prices of Swiss real estate mutual fund units are market values and not appraised values, but the market values will be quite close to net asset values. The returns on such funds are, thus, influenced by the valuations of the properties and also by stock market factors (see Hoesli and Anderson, 1991; and Hoesli, 1993).

Swiss real estate mutual funds had a mean return of 7.19% and a standard deviation of 10.19% over the period 1961–97, compared to a mean return and a standard deviation of 8.62% and 21.89%, respectively, for shares, and of 5.05% and 4.58%, respectively, for bonds. Swiss real estate mutual funds, thus, exhibit a lower risk and return than shares and a higher risk and return than bonds.

Correlation with shares and bonds[19]

High degrees of correlation have been found between the returns on US REITs and shares. Brueggeman and Fisher (1997) report a correlation coefficient of 0.6912 between equity REITs and shares for the period 1978–94. When REIT returns are ungeared, however, the correlation coefficient between REIT returns and stock returns decreases: Fisher *et al.* (1994), for instance, report a correlation coefficient of 0.44 between an ungeared REIT index and the returns on shares for the period 1978–92. Brueggeman and Fisher (1997) also report the correlation coefficient between equity REITs and corporate bonds for the period 1978–94. This correlation is 0.4151, which is almost the same coefficient as between shares and bonds (0.3925).

Equity REITs also exhibit a low correlation with direct property returns (as measured by the unsmoothed NPI index). Corgel *et al.* (1998) report a correlation of 0.090 between equity REITs and direct property for the period from 1978 to the second quarter of 1995 (the correlation between direct property and stocks is −0.072).

Khoo *et al.* (1993) report declining correlation coefficients between US REITs and common stocks in the 1980s. They suggest that, over time, security analysts have become more equipped to reveal the value of the property behind the securities. A similar result of declining correlation coefficients between REITs and common stocks is found by Ghosh *et al.* (1996). The correlation coefficient with monthly data was 0.77 for 1985–7, 0.71 for 1988–90, 0.38 for 1991–3 and 0.40 for 1994–June 1996.

For France, Hoesli and Thion (1994) report a correlation coefficient between property stocks and shares of 0.55 for the period 1982–92. The correlation between property stocks and bonds is 0.47 (see also Thion and Riva, 1997). In Switzerland, the correlation coefficient between property companies and stocks is 0.488 and the correlation between property companies and bonds is 0.635 for the period 1961–97.

High degrees of correlation between the returns on property companies and those on shares are found for most countries by Eichholtz (1997b), with the exception of Austria and Germany and, to a lesser extent, Belgium, the Netherlands and Switzerland (see Table 10.9). Eichholtz (1997b) also reports these correlation coefficients for three regions (Europe, North America and the Far East) and the world.

Efficient sets

The return and risk characteristics of US REITs are quite similar to those of common stocks. Further, the correlation coefficient between REIT returns and share returns is relatively high. Even with such parameters REITs have a positive role to play in diversifying multi-asset portfolios. Mueller *et al.* (1994), for instance, examine the benefits of including REITs in multi-asset portfolios for the period from January 1976 to June 1993. Adding REITs to the portfolio would have yielded an additional historical return of between 1 and 14 basis points per month, or an annual increase in returns of between 12 and 168 basis points (without compounding), for the same level of risk. The weight of REITs in

Table 10.9► Correlation coefficient between the return on property companies and the return on common stocks for 19 countries, three regions and the world (January 1987–December 1996)

Country	Correlation coefficient
Austria	0.12
Belgium	0.38
France	0.61
Germany	0.13
Italy	0.64
Netherlands	0.36
Norway	0.43
Spain	0.53
Sweden	0.59
Switzerland	0.35
UK	0.80
Canada	0.56
USA	0.72
Australia	0.76
Hong Kong	0.96
Japan	0.83
Malaysia	0.80
New Zealand	0.56
Singapore	0.93
Europe	0.82
North America	0.71
Far East	0.79
World	0.85

Source: Eichholtz (1997b).

optimal portfolios is large (up to 80% in some portfolios). For the period from the first quarter of 1982 to the fourth quarter of 1993, Liang *et al.* (1996) report weights in the 15–20% range for equity and apartment REITs.

Hoesli and Thion (1994) construct optimal portfolios for France for the period 1982–92. They find that the weight of property shares is in the 10–38% range depending on the level of risk aversion. The results for Switzerland should prove of interest as Swiss real estate mutual funds have been shown to behave quite differently from common stocks. Figure 10.1 contains, for Switzerland, the efficient frontier which contains combinations of shares and bonds and the efficient frontier which contains property shares as well as shares and bonds for the period 1961–97. As can be seen, the benefits of including real estate mutual fund units in portfolios of shares and bonds are quite large. For an average return of 6.24%, for instance, the risk would fall from 8.98% to 7.05% if real estate mutual fund units were included in the portfolio.

Empirical evidence for Switzerland suggests that both direct property and shares of property companies should be included in multi-asset portfolios (see Hoesli and Hamelink, 1996, 1997). Of the two types of property, direct property

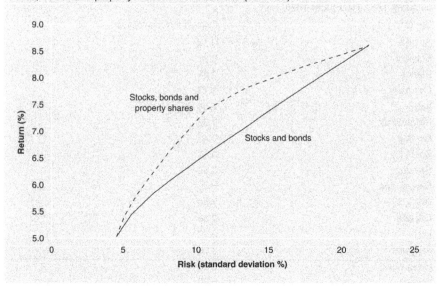

Figure 10.1► Efficient frontier containing stocks and bonds and efficient frontier containing stocks, bonds and property shares for Switzerland (1961–97)

provides the greater diversification benefits. Thus, if one type of property has to be selected, the emphasis should be placed on direct holdings, but shares of property companies provide diversification benefits even when direct property holdings are already included in the portfolio. For smaller investors, shares of property companies may constitute the only means of investing in the property market (see Chapter 11).

10.5 Reconciling suggested and actual weights

The previous sections contain reviews of studies which have analysed the role of commercial property and residential property in diversifying multi-asset portfolios. The impact of including the shares of property companies is also examined. These studies are usually undertaken in a Modern Portfolio Theory (MPT) framework. The general consensus of these studies is that a significant proportion of an investment portfolio should be allocated to property. The suggested weights for property are in the 15–20% range, whether commercial or residential property is considered.

When the portfolios of institutions are analysed, it is striking that the actual weights allocated to property in these portfolios are, in most countries, lower – and in some instances much lower – than the weights suggested in the financial economics literature. Dumortier (1997) reports weights for property in the 2–13% range (see Table 10.10). For Switzerland, the weight allocated to property by pension funds is 16%.

The weight allocated to property in institutional portfolios is often inversely related to the percentage of owner-occupants in the residential sector, which is not

Table 10.10► Weight allocated to property by institutions in selected countries[20]

Country	Weight in property
USA	3%
UK	6%
Netherlands	13%
Japan	7%
Canada	2%
Australia	11%
Denmark	10%
Sweden	12%
France	7%

Source: Dumortier (1997).

Table 10.11► Property holdings of life insurance companies in selected countries (1988 and 1996)

Country	Property allocation	
	1988	1996
USA	8.15%	2.59%
UK	17.22%	7.23%
Sweden	—	4.81%
Austria	14.37%	9.86%
Netherlands	7.20%	5.00%
Italy	—	14.57%
France	6.02%	12.32%
Spain	28.81%	16.58%
Belgium	6.39%	4.57%
Australia	0.00%	6.95%

Source: Chun and Shilling (1998).

surprising given that the greater the ownership, the fewer the apartments in which to invest. In Switzerland, for instance, only 31% of households own their home. This percentage amounts, for instance, to 56% in France, 67% in the UK, 68% in Italy, 75% in Spain and 79% in Greece (see Hoesli, 1998). From the perspective of individuals, there might not be a very big difference between suggested and actual property holdings. In some countries, individuals gain exposure to property by purchasing a home, while in others this exposure is gained indirectly through the holdings of their pension fund.

Detailed survey results of Dutch pension funds and insurance companies are presented by De Wit (1996) and of US pension funds by Worzala and Bajtelsmit (1997). The allocations vary by fund size: generally the smaller the fund the smaller the property holding. The weight allocated to property has declined in several countries in recent years. Chun and Shilling (1998) report declining property allocations between 1988 and 1996 for life insurance companies in several countries (see Table 10.11). In France and Australia, however, the allocation to property has increased.

Several reasons have been given to explain the differences between the suggested and the observed allocations to property by institutions (see, for instance, MacGregor and Nanthakumaran, 1992; and Ball *et al.*, 1998). These include limitations of the data used in the MPT analyses, the illiquidity of property, transaction costs, the size of the fund, omitted assets, competitor risk and liabilities. Each of these possible reasons is discussed below.

Limitations of data

The vast majority of the studies which have analysed the benefits from including property in a mixed-asset portfolio have made use of *ex post* data (an exception to this is the study by Lee *et al.*, 1996), whereas portfolio construction is undertaken with *ex ante* data (forecasts). Mei and Saunders (1997), for instance, report a negative correlation between current and past property returns and their future expected values: unexpectedly large excess returns today are associated with smaller future expected excess returns. However, institutions have long time horizons and their strategic allocations should not be very different from studies which have made use of historical data over a long time period.

Property data are far from being optimal. For example, appraisal-based returns are smoothed. However, even when the series have been desmoothed, the MPT analyses suggest significant allocations to property. More sophisticated methods exist to construct property indices, such as the hedonic method and the repeat sales method. These methods are now quite widely used in the residential sector and are developing in the commercial property sector. The empirical studies examined in this chapter encompass a wide variety of methods and countries, so that the data limitation issue should not represent too much of a problem.

Illiquidity

In MPT analyses, risk is measured by the standard deviation of returns. One important source of risk, that is liquidity risk, is not captured by the standard deviation, and this will lead to the role of property in multi-asset portfolios being overestimated. When direct property data are used, illiquidity and its associated risk should in all cases be taken into account. Studies which have done this include MacGregor and Nanthakumaran (1992), Leggett (1995) and Hoesli and Hamelink (1996). These authors show, however, that even when the illiquidity of property is accounted for, property has a substantial role to play in diversifying multi-asset portfolios. It might simply be that investors overestimate the negative impact of property's illiquidity when making strategic allocation decisions.

Transaction costs

An important factor which can help explain the differences between observed allocations and the suggested allocation are transaction costs. The MPT analyses usually ignore transaction costs which are much higher in the property market. As reported in Chapter 12, transaction costs for commercial property are as high as 9% in Japan, 10% in Portugal and 18.4% in France. The more active the fund management, the greater these costs and hence the lower the returns. Even in countries where such costs are lower, they remain substantially higher than is the case for shares.

Size of the fund

Systematic risk can only be diversified away in large property portfolios (see Chapter 9). Smaller investors are, thus, subject to higher risk and should not hold property. The MPT results should be compared to the allocations of larger funds only. When this is done, the discrepancy between theoretical and actual allocations is not as pronounced.

Omitted assets

The role of property is usually examined for portfolios which contain shares, bonds and, in some cases, cash. All asset categories should be included in such analyses to gain better insight as to the benefits of considering property as an investment alternative. In this context, one asset category which has seen its allocations increase over the years is that of overseas shares. Gordon *et al.* (1998), for instance, show that the efficient frontier which contains US shares, bonds and REITs is dominated by the frontier which contains domestic and international stocks, which suggests that international stock diversification would be more effective than domestic inter-asset diversification through REITs.

It can be argued that, if international stocks are considered, international property investments should be considered too, and that in this context an optimal allocation of 15–20% for property might again prevail. The obvious counterargument is that an international property strategy is difficult to undertake as it requires, among other things, local market knowledge. This counterargument is not valid, however, if international property shares are considered. In fact, Eichholtz (1996) shows that the average correlation of property share returns across countries is lower than the equivalent figures for shares and bonds (see Chapter 12). This implies that international diversification can reduce the risk of indirect property portfolios even more than it can reduce the risk of common stock and bond portfolios. Gordon *et al.* (1998) show that an efficient frontier which contains domestic stocks, international shares and international property shares dominates an all shares frontier, making the point that international property share diversification is very effective.

Competitor risk

Portfolio strategies are typically developed in relation to a benchmark against which fund performance is measured. This may be a market benchmark or competitors in the same business area. Funds will choose allocations which do not differ very much from those of competitors as poor relative performance is likely to mean loss of business. The conventional benchmarks comprise both large and small funds but the latter tend not to hold property as they cannot construct diversified portfolios. Accordingly, the benchmarks have low weights allocated to property. These are further reduced if an unweighted average is used to construct the benchmark. Thus, funds which consider tracking error risk are likely to hold low weights in property (see Chapter 9). This is an example of the principal–agent problem: the interests and objectives of the principal (the investor) are not necessarily coincident with those of the agent (the fund manager) (see Ball *et al.*, 1998).

Liabilities

Institutional investors must also consider their liabilities, and must hold assets which are likely to allow them to meet such liabilities. Not being able to meet liabilities represents an important additional dimension of risk. Accordingly, investments are not made purely on the basis of the annual return and volatility of assets, but by taking account of the funds' liabilities. One strategy would be to match the durations of the assets and liabilities (see Chapter 7).

Differences in liabilities help to explain fund structures (see Chapter 2). As liabilities are often linked to inflation, assets which provide protection against inflation are often sought. Hamelink *et al.* (1997) examine the inflation protection effectiveness of shares, bonds, property and property shares in the USA and the UK for the period 1978–95. When the 95% probability of achieving a positive real return is computed for the various asset classes, property provides the best protection in the USA, while the level of protection is highest with shares in the UK. In the USA, seven years are needed with property to have a 95% probability of a positive real return. For property shares, nine years are needed, while 12 years are needed for bonds and 16 years for stocks. In the UK, four years are needed for shares, 13 years for property, 19 years for bonds and 37 years for property shares. Little evidence has been found, however, to support the contention that there is full compensation for expected and unexpected inflation on a period-to-period basis (see Hoesli, MacGregor, Matysiak and Nanthakumaran, 1997).[21]

10.6 Summary and conclusions

In this chapter, the role of property in mixed-asset portfolios has been examined. Exposure to property can be either direct (in the commercial or residential sector) or indirect (by the purchase of shares of property companies). In order to construct efficient frontiers, the average return and risk for each asset class and the correlation coefficients between the returns on each pair of assets are needed. Each of these parameters was examined for commercial, residential and indirect property. In particular, the return and risk characteristics of property were compared with those of shares and bonds.

When direct property is considered, either appraisal-based indices or transaction-based indices can be used. Appraisal-based indices are usually used for commercial property, while transaction-based series are more common for residential property. In terms of comparison of the return and risk parameters of property to those of shares and bonds, the results are generally similar whether commercial or residential property is considered. These results suggest that property yields a higher return and risk than bonds and a lower return and risk than shares. When property shares are considered, the return and risk parameters are quite close to those of shares in most countries.

Direct property returns have been shown to be lowly correlated with those on shares and bonds. The correlation between property and shares is often slightly positive, while the correlation between property and bonds is in most cases slightly negative. In contrast, indirect property returns exhibit high levels of correlation with stock returns. In all cases, property is shown in this chapter to constitute an attractive portfolio diversifier which enables a reduction in risk (for a given return) or an increase in return (for a given risk level). Several authors have suggested an optimum weight of 15–20% for property in multi-asset portfolios.

This suggested weight lies above the actual property allocation of institutions in most countries. For instance, UK institutions hold on average 6% in property, while US institutions hold on average 3% in this asset class. Several potential reasons were given which could explain the difference between suggested and actual allocations to property. These include the limitations of property data, the illiquity of property, the high transaction costs for property, the size of the fund, the omission of other asset classes in the Modern Portfolio Theory (MPT) analyses, competitor risk and the matching of assets with liabilities.

The limitations of the data and the illiquidity of property can be taken into account in the MPT analyses, so they should not lead to serious biases in the theoretical weights for property. The other reasons certainly provide an explanation for the difference between theoretical and suggested weights. In particular, the actual allocations reported for institutions represent averages which include small funds with no property. As the allocation to property usually rises with the size of the fund, the discrepancy between theoretical and actual weights should be less pronounced for larger funds.

It was also noted that the weight allocated to property in institutional portfolios is often inversely related to the percentage of owner-occupants in the residential sector. In Switzerland, for instance, only 31% of households own their home, whereas this percentage is 67% in the UK and 79% in Greece. Thus, from the perspective of individuals, there might not be a very big difference between suggested and actual property holdings. In some countries, individuals gain exposure to property by purchasing a home, while in others this exposure is gained indirectly through the holdings of their pension fund.

Notes

1. The results from empirical studies are reported with the number of decimal points which are reported in these studies. This number varies from one study to another.
2. From a theoretical perspective, this does not necessarily constitute a problem. Mean-variance dominance is not incompatible with a CAPM world, since the studies reviewed in this chapter focus on total risk rather than systematic risk (see Chapter 7). As mentioned by Liu et al. (1995), '. . . all assets could still plot on the security market line in equilibrium, given a CAPM world, regardless of whether one asset (portfolio) dominates another asset (portfolio) from a mean-variance perspective'.
3. The figures for South Africa relate to capital returns only and not to total returns.
4. The return frequencies reported in this table are the frequencies of returns which are used to calculate the factors between 'true' standard deviations and standard deviations of smoothed returns. This frequency is not always the same as that used for the

unsmoothing procedure. For this procedure, Fisher *et al.* (1994) use quarterly returns and Newell and MacFarlane (1996) use monthly returns.

5. The smoothing parameter used to unsmooth an index presumably also changes over time. On this issue, see Matysiak and Wang (1995) and Chaplin (1997).
6. When shorter time periods are considered (1985 to the second quarter of 1995, and 1990 to the second quarter of 1995), however, property appears to be dominated by bonds. This is due to the strong impact of the property market slump in the early 1990s on shorter time periods.
7. An early attempt to construct a hedonic index for commercial property is made by Hoag (1980). He constructs both an equally weighted index and a value-weighted index of US industrial property for the period 1973–8. The mean return of property is 8.7% and the standard deviation is 19.1% when the equally weighted index is used, and 14.2% and 17.2%, respectively, when the value-weighted index is used. For shares, the mean return is 3.7% and the standard deviation 20.8%, whereas for bonds the figures are 6.3% and 8.0%, respectively. Industrial property, thus, appears to yield a higher return and risk than bonds. Property also appears to have a higher return and a lower risk than shares, but the comparison with shares is problematic as the stock market was bearish during the short period examined by Hoag.
8. Another study for the UK is undertaken by Newell and Brown (1994). For the period January 1988 to December 1992, Newell and Brown de-smooth the IPD index of monthly returns. The correlation between property and shares is −0.173 (−0.015 with the smoothed series), and the correlation between property and long-term bonds is 0.004 (−0.136 with the smoothed series).
9. Hedonic indices of commercial property have also been used in the USA by Hoag (1980) and Cole (1988). Hoag (1980) computes the correlation between property returns and those on shares and bonds. He reports a correlation of −0.07 between property returns and share returns, and a correlation of −0.31 between property returns and bond returns. His results, however, are probably not representative of the true degree of correlation between property returns and the returns on shares and bonds. First, the period examined is very short: from 1973 to 1978. Second, his hedonic index is constructed on the basis of industrial properties only. Cole (1988) builds a hedonic index based on properties which are well diversified. He reports a correlation of −0.055 between property and shares, and a correlation of −0.038 between property and bonds for the period 1982–6. Both correlation coefficients are statistically not different from zero.
10. No attempts have been made to investigate the role of commercial property in diversifying mixed-asset portfolios using transaction-based indices. The results concerning the impact of considering commercial property as a portfolio diversifier thus only refer to appraisal-based indices.
11. MacGregor and Nanthakumaran (1992) calculate the minimum and maximum returns on the efficient frontier. The range of returns is then divided to produce ten portfolios with the same increment in return between successive portfolios. They only report the composition of portfolios with second lowest and second highest return.
12. Newell and MacFarlane (1995) suggest an adjustment factor of approximately 2, but do not construct efficient frontiers.
13. The analysis by Webb and Rubens (1987) is not limited to commercial property as these authors consider residential real estate, business real estate and farmland.
14. A weight of 9% only for property is suggested by Kallberg *et al.* (1996).
15. For shares, bonds and commercial property, the period is 1969–87.
16. It must be remembered that the commercial property index includes some residential property, which biases the correlation coefficient upwards.

17. A small-value share (or small-capitalisation share) is a share of a company whose market value is small.
18. A REIT is a specialised form of business trust that owns property or investments in property (see Chapter 11).
19. The correlations between indirect property and shares are higher than between direct property and shares. This may be explained by the process of price discovery. Information is incorporated more quickly into price in the indirect market with the result that, although there are low contemporaneous correlations between the direct and indirect markets, the lagged correlations are much higher. For a fuller discussion see Chapter 11.
20. The weight allocated to property is a simple arithmetic average (unweighted average) in some countries, while it is a weighted average in other countries. As large institutions usually hold a higher proportion of their assets in property, the aggregate figure of the property holdings of institutions will be higher in countries where a weighted average is used.
21. Whether asset returns are correlated with inflation (and its expected and unexpected components) is known as inflation hedging.

Further reading

Ball, M., Lizieri, C. and MacGregor, B. D. (1998) *The economics of commercial property markets*, Routledge, London: Chapter 11.

Brueggeman, W. B. and Fisher, J. D. (1997) *Real estate finance and investments*, Irwin, Chicago (tenth edition): Chapter 21.

Coleman, M., Hudson-Wilson, S. and Webb, J. R. (1994) Real estate in the multi-asset portfolio, in *Managing real estate portfolios*, Hudson-Wilson, S. and Wurtzebach, C. H. (Eds), Irwin, Burr Ridge (IL), 98–123: Chapter 3.

Corgel, J. B., McIntosh, W. and Ott, S. H. (1995) Real estate investment trusts: a review of the financial economics literature, *Journal of Real Estate Literature*, 3(1), 13–43.

Corgel, J. B., Smith, H. C. and Ling, D. C. (1998) *Real estate perspectives: an introduction to real estate*, Irwin/McGraw-Hill, Boston (third edition): Chapter 7.

Leggett, D. N. (1995) An empirical analysis of efficient real property liquidity premiums, in *Alternative ideas in real estate investment*, Schwarz, A. L. and Kapplin, S. D. (Eds), sponsored by the American Real Estate Society, Kluwer Academic Publishers, Boston, 113–27: Chapter 7.

Liu, C. H., Grissom, T. V. and Hartzell, D. J. (1995) Superior real estate performance: enigma or illusion? A critical review of the literature, in *Alternative ideas in real estate investment*, Schwarz, A. L. and Kapplin, S. D. (Eds), sponsored by the American Real Estate Society, Kluwer Academic Publishers, Boston, 59–82: Chapter 4.

MacGregor, B. D. and Nanthakumaran, N. (1992) The allocation to property in the multi-asset portfolio: the evidence and theory reconsidered, *Journal of Property Research*, 9(1), 5–32.

Pagliari, J. L., Jr and Webb, J. R. (1995) Strategic asset allocation: a comparative approach to the role of real estate in a mixed-asset portfolio, in *The handbook of real estate portfolio management*, Pagliari, J. L. (Ed.), Irwin, Burr Ridge (IL), 1059–111: Chapter 26.

Chapter 11

Alternative forms of investment in property

11.1 Introduction

In Chapter 10 it was shown that direct property investments are good portfolio diversifiers as their return and risk characteristics differ from those of financial assets, and as their returns are lowly (negatively or positively) correlated with those on financial assets. For large institutions, it appears that direct property investments should be included in the portfolio to achieve optimal portfolio diversification.

Some characteristics of direct property investments, however, may limit the attractiveness of such investments. For smaller institutions, for instance, the large lot size of direct property investments will not enable them to build well-diversified property portfolios. The property component of the portfolio will thus exhibit high levels of non-systematic risk. Also, the illiquidity of direct property may discourage some investors from considering such investments. Other characteristics such as high transaction costs or the need for local market knowledge may also constitute a barrier to entering the direct property market (these issues are considered in more detail in Chapter 2).

In order to alleviate the drawbacks of direct property investments, indirect property vehicles have been developed. The basic idea of these indirect vehicles is to pool property assets. By purchasing shares of a pool of property assets, an investor can indirectly hold property assets. Far better levels of within property portfolio diversification can be achieved as the owner of property companies' shares will hold a small proportion of a large property portfolio. As will be seen in this chapter, some indirect vehicles predominantly have a primary market, whereas some also have an active secondary market. An important issue which will be addressed in this chapter is whether such indirect vehicles reflect the underlying property assets. Ownership of one property can also be split up between several investors: this is known as securitisation.

Indirect property vehicles clearly represent an alternative to direct property investments. Other alternatives exist to direct ownership of property. Instead of taking an equity position in a building or in a group of buildings, an investor can lend money to someone who wants to construct or purchase a property. Such vehicles are debt instruments, as opposed to equity instruments. Mortgages are an example of a debt instrument whereby the lender receives interest payments as specified in the loan agreement. The underlying property provides security for the loan. With a conventional loan the lender does not participate in any capital growth in the property asset but may be affected by default if property values fall substantially. In the last ten years, debt instruments have been developed which enable the lender to participate in growth, making these hybrid debt-equity investments.

Investors can also lend money indirectly by purchasing shares of companies which lend money to property developers or property investors. Various ways of achieving this objective exist, in particular in the USA. For example, mortgage-backed securities have been developed both for residential and commercial properties. An exposure to the property market can, thus, be achieved by taking an equity or a debt position either directly or indirectly.

Another way of gaining exposure to the property market is through derivative vehicles. Derivatives are instruments whose returns are derived from those of other (underlying) investment vehicles. Futures and options are examples of derivative vehicles. Such instruments can be used as hedging instruments or as investments with high levels of gearing. Attempts to establish a futures market in property in London have failed. Recently, a number of index-based derivatives have been launched in London.

This chapter is organised as follows. Indirect equity instruments are presented first: property company shares in Section 11.2, other collective equity investment vehicles in Section 11.3 and unitisation in Section 11.4. In Section 11.5, direct and indirect debt instruments are discussed. Section 11.6 contains a discussion of property derivatives, while Section 11.7 contains a summary and conclusions.

11.2 Property company shares

Direct property investments exhibit adverse features which can in some cases dissuade investors from considering this class of asset when constructing multi-asset portfolios. These features include the lack of a central market for property, the illiquidity of property investments, transaction costs, the high unit value of such investments and the management requirements of property (see Chapters 2 and 10).

As a result of these problems, various types of indirect property vehicles have been developed. These include property company shares which bear different names according to the country, for instance Real Estate Investment Trusts (REITs) in the USA and real estate mutual funds in Switzerland. Such companies usually invest in income-producing properties and there is a secondary market for their shares. Property company shares are discussed in this section.

Indirect property vehicles also exist for which the share price is set according to valuation of the underlying properties. The Property Unit Trusts (PUTs) in the

UK and Commingled Real Estate Funds (CREFs) in the USA are such vehicles. The secondary market for such vehicles is limited. Securitised single-asset vehicles also exist, whereby an investor acquires an interest in individual properties. These vehicles are reviewed in Sections 11.3 and 11.4, respectively.

UK property companies

Unlike US REITs, UK property companies have no particular legal form or any specific tax advantage, but are ordinary public companies. There are broadly two types of property companies. The first type, the property investment companies, acquire or develop properties and then retain them. In other words, they invest in income-producing properties. The rental income is used to cover running costs and interest charges and to pay dividends. The company's market price should reflect the value of the underlying properties. Whether this is true or not is discussed below. The second type of firm, developers-traders, construct or acquire properties, then sell them. The difference between sale price and purchase price (or development costs) from those sales provide company profits. As a result, such companies' shares tend to be priced according to their expected earnings growth.

As pointed out by Barkham (1995), 'the distinction between investors and developers-traders is not clear-cut because most traders own investment property and most investors trade. Furthermore, both investors and traders engage in property development – investors for long-term holdings and traders for immediate resale.' Some property companies specialise in one property type or one region, whereas some companies hold portfolios diversified both by property type and geographically. However, most holdings are concentrated in the South East (around London). Land Securities, for instance, the largest property company, is strongly oriented towards central London property.

It is estimated that 43% of Land Securities assets are in London, 21% in the rest of the South East, 19% in the rest of the UK and 17% in international assets (figures by UBS Phillips and Drew, reported by Barkham and Geltner, 1995). Compared to the Investment Property Databank (IPD) index (see Chapter 4), this represents a larger weight in London, approximately the same weight in the rest of the South East and a smaller weight in the rest of the UK. Property companies in the UK vary considerably in levels of gearing.

The Financial Times–Stock Exchange Property Sector index encompasses approximately 95% of the market capitalisation of property companies. In mid-1995, the market capitalisation of companies included in the FTA Property Sector index was £16.1bn. The average annual total return for property companies ranged from a low of –23% in 1991 to a high of 96% in 1994 over the period 1985–94 (Barkham, 1995). The returns on such companies are, thus, quite volatile (see Chapter 10).

The concept of Net Asset Value (NAV) is widely used for UK property company shares. NAV is the total value of the assets of a company (as determined by valuation of the properties) minus the value of its liabilities. In general, property company shares trade at a discount to NAV, and the average discount to NAV varies over time as does the discount of individual companies. In both cases, a premium sometimes occurs rather than a discount. Possible reasons for the discount to NAV include asset management fees, taxation effects, agency costs, excess volatility from stock market 'noise' because of uninformed trading in shares and the stock market's lack of confidence in the management of the property company (Barkham, 1995; Barkham and Ward, 1999).

US REITs

A Real Estate Investment Trust (REIT) is a real estate company or trust that has elected to qualify under certain tax provisions to become a pass-through entity that distributes to its shareholders almost all of its earnings and capital gains generated from the disposition of its properties. The real estate investment trust does not pay taxes on its earnings, but the distributed earnings do represent dividend income to its shareholders and are taxed accordingly. To qualify as a REIT for tax purposes, the trust must satisfy certain asset, income and distribution requirements. These include:

- at least 75% of the value of a REIT's assets must consist of real estate assets, cash, and government securities
- at least 75% of gross income must be derived from rents, interest on obligations secured by mortgages, gains from the sale of certain assets, or income attributable to investments in other REITs
- distributions to shareholders must equal or exceed the sum of 95% of REIT taxable income.

The three principal types of REIT are equity REITs (EREITs), mortgage REITs (MREITs) and hybrid REITs (HREITs). The National Association of Real Estate Investment Trusts (NAREIT) uses a 75%-of-assets cutoff. If a trust holds more than 75% of its assets in equity real estate, for instance, it is an equity REIT. EREITs constitute an indirect way of investing in equity real estate, while MREITs are a way of lending funds to property owners or developers. MREITs are discussed in Section 11.5.

The classification of REITs according to the mix of assets is the most obvious way of proceeding. As reported by Downs and Hartzell (1995), however, REITs can be classified along many other different dimensions such as:

- Property type: some REITs invest in several types of properties, whereas others focus on one sector.
- Health care REITs: these specialise in health care facilities such as hospitals, medical office buildings and assisted living centres. Health REITs are usually not considered as being real estate.
- Geographic concentration: some REITs invest in properties across the USA, whereas most REITs have some geographic focus.
- Operating time frame: some REITs are formed to operate indefinitely (perpetual REITs), and some are formed for a prespecified period of time (finite-life REITs).
- Closed versus open-ended REITs: closed REITs specify a maximum amount that the trust will raise for investment, whereas open-ended REITs create and sell new shares as they find new investment opportunities. As noted by Corgel *et al.* (1998), 'this ability to issue additional equity can enhance the ability of the

Table 11.1► Types of REITs in volume and number (as of 31 January 1999)

Type	Value	Value (%)	Number	Number (%)
Equity	$128.8bn	93%	176	83%
Mortgage	$6.2bn	5%	25	12%
Hybrid	$2.8bn	2%	10	5%
Total	$137.8bn	100%	211	100%

Source: National Association of Real Estate Investment Trusts (NAREIT).

REIT management to make timely property acquisitions, especially given the REIT's limited ability to finance acquisitions with retained earning because they are required to pay out 95% of taxable income in dividends.'

Most REITs are equity, infinite life and open-ended. A new distinction is used between 'old' and 'new' REITs. Old REITs are the REITs which were created prior to 1990, and new REITs are those created post-1990. Pre-1990 REITs were largely passive investment vehicles that owned diverse portfolios of properties. Most REITs arranged for both portfolio and property management through an external adviser. Many pre-1990 REITs were also finite-life REITs, which limited their growth potential.

The post-1990s equity REITs differ from their predecessors in their organisation, business plans and ownership structure. The managements of the new REITs usually have substantial equity positions in the company and all are infinite-life REITs. Property management is either done internally or by management that works solely for one REIT. Property investments are focused by property type and geographic market. The new REITs have attracted more institutional investors than their predecessors. As of 1994, 96 REITs were 'old' REITs and 127 were 'new' REITs. Ninety-three per cent of the new REITs were EREITs, 5% were MREITs and 2% were HREITs.

From 1972 to the beginning of 1999, the total market capitalisation of the REIT industry grew from about $1.8bn to $137.8bn. Between 1990 and the beginning of 1999, the number of REITs grew rapidly from 96 to 211. As can be seen in Table 11.1, 93% of REITs are equity REITs. One hundred and sixty-seven REITs ($131.1bn) are quoted on the NYSE, 27 REITs ($5.3bn) on the AMEX and 17 REITs ($1.4bn) on the Nasdaq.

Three indices of securitised property are used in the USA: the NAREIT index, the Wilshire index and the Lehman Brothers index. As of the end of 1994, 226 securities were included in the NAREIT index, 119 in the Wilshire index and 100 in the Lehman index. Various subindices of the NAREIT index are available for EREITs, EREITs exclusive of healthcare REITs, MREITs, HREITs and by property type (for more details, see Giliberto and Sidoroff, 1995).

Property companies in other countries

Property companies exist in most developed countries. They exhibit a broad diversity. First, property companies encompass investment companies and developers-traders. Second, in some countries they invest exclusively domestically (such as in Switzerland), whereas in some countries selected funds invest internationally as well as nationally (such as in the Netherlands). Third, in some countries, property companies are tax exempt (for example, the real estate investment funds in Germany), whereas they are not in other countries (for example, the Swiss real estate mutual funds). Fourth, property companies can invest in housing, in commercial property or in both. Property companies in Switzerland, for example, invest predominantly in housing, while UK property companies mostly focus on commercial properties. Finally, the level of gearing (leverage) varies from one country to another and sometimes within a country as well. Property companies in Japan are highly geared, but Swiss real estate mutual funds are not.

Eichholtz and Koedijk (1996b) report that the number of equity property companies in the world grew from 124 as of the end of 1983 to 431 as of the end of 1995. Twenty-nine per cent of these companies were in North America, 39% in Europe and 32% in the Asia–Pacific region. Of the 431 companies, 191 (44%) were diversified (that is, they do not have more than 60% of their assets invested in one property type), 68 (16%) invested predominantly in residential property, 79 (18%) in retail property and 49 (11%) in offices. The bulk of these companies invest more than 75% of their funds in their own country. Twenty-two companies invested between 25% and 75% of their assets abroad, and 16 companies held more than 75% of their assets abroad.

As discussed in Chapter 4, international property securities' indices exist. These include the international property share index of the Datastream Global Indices, the GPR indices, the Morgan Stanley Capital International property share index and the Salomon Brothers World Equity Index-property. Subindices by country are also available and national real estate securities' indices also exist in several countries (for instance, the SBV index for the Netherlands and the Bopp ISB A.G. index for Switzerland).

The GPR-LIFE Global Real Estate Securities index is the most comprehensive international real estate securities' index. As of the end of September 1998, it encompassed 415 securities in 27 countries with a total market capitalisation of $306bn. Country and regional indices are also available. Developers are excluded from the index. In comparing property securities across countries, it is important that the same definition of a property company be used. In this respect, the GPR indices are the most appropriate.

Advantages and disadvantages of property company shares

Property company shares do not exhibit the disadvantages of direct property investments. First, they are more liquid investments, which means that the time taken to make trades is far shorter than in the direct market. Second, as unit cost is low, part of a portfolio can be allocated to property even if the portfolio is small. This would not be possible with direct investments. Third, good levels of within property portfolio diversification can be achieved by buying shares of a few property companies, or even of just one company provided that its assets are well diversified. Finally, since the shares are publicly traded, the share price is known at any time, which is not the case with direct property investments.

Property company share ownership also entails disadvantages. In some countries such as the UK and Switzerland, there is no full tax transparency as property companies are taxed on their profits. This is a major disadvantage for tax-exempt investors such as pension funds who cannot claim back corporation tax. Also, there is often a lack of management control, and the shareholder may find it difficult in some countries to obtain full information on the property assets and the development schemes of the company.

A critical issue which has been the focus of much attention in the literature is whether these property companies behave like direct property or whether they are another type of stock. As shown in Chapter 10, the returns on such shares are volatile, which would suggest that they behave more like stocks than like appraisal-based property indices. Several studies have been conducted for US equity REITs.

It is usually shown that the income return component of US REITs is correlated with that of direct property investments, but that the capital return component is highly correlated with that of common stocks. Giliberto and Mengden (1996), for instance, report a strong and positive correlation between securitised and unsecuritised property cash flows. This relationship does not exist when the capital return components are considered. Giliberto and Mengden suggest that differences between the securitised and unsecuritised markets obscure the positive relationship between the cash flows of both assets.

As outlined in Chapter 10, a study by Khoo et al. (1993) reports declining correlation coefficients between US REITs and common stocks in the 1980s. This may be because, over time, security analysts have become better equipped to reveal the value of the property behind the securities. Other studies also point to REITs now behaving a little more like direct property than used to be the case. Ghosh et al. (1996) report two other facts which may indicate that property company shares are becoming more like property and less like stocks. First, REITs appear to be less liquid than other comparable-size stocks. Second, institutional investment in REITs is growing, although it is still below the level of comparable stocks.

High levels of correlation coefficients have been reported for other countries also. Eichholtz and Hartzell (1996), for instance, report a strong contemporaneous relationship between property shares and the stock market on which they trade for the USA, the UK and Canada. Eichholtz (1997b) computes correlation coefficients between property company shares and common stocks in 19 countries, three regions and the world. The country correlation coefficients range from a low of 0.12 for Austria to a high of 0.96 for Hong Kong. For Europe, the correlation is 0.82, for North America 0.71 and for the Asia–Pacific region 0.79. For the world, the correlation amounts to 0.85 (see also Chapter 4).

The relationship between the direct and indirect markets is also considered in the literature on price discovery – that is, the process by which asset prices are formed. The opinions of market participants are aggregated into a single statistic, the market price. Although they are traded in separate markets, and factors such as gearing affect values in the indirect market, both direct and indirect property have values which are related through the common component in value, the ownership of properties. In these circumstances, price discovery of the common component may occur first in one market and then be transmitted to the other market. Barkham and Geltner (1995) investigate the UK and US property markets and report that price discovery of the common commercial property value element occurs first in the indirect market and then later in the direct market.[1] The lag between the markets is less, and information transfer stronger, in the UK than in the USA, perhaps because of the greater homogeneity of the property market in the UK and because a larger proportion of properties has been securitised. This means that the lagged correlations between the direct and indirect markets are greater in both markets than the contemporaneous correlations.

11.3 Other collective equity investment vehicles

In the UK and the USA, other collective equity investment vehicles are available. Such vehicles are mainly transacted on the primary market, although there is a limited secondary market, and the price of the units is based on the regular valuations of the properties.[2] In the UK, institutions may invest in property through Property Unit Trusts (PUTs) and Managed Funds. PUTs are pools of property investments held in the name of a trust. Most trusts are open-ended. Managed Funds are similar in objective to PUTs. They consist of unitised property funds managed for occupational pension schemes by insurance companies. Both PUTs and Managed Funds confer tax advantages on tax-exempt investors (pension funds and charities). Since 1991, Authorised PUTs have been permitted which may be advertised to the general public.

A similar vehicle exists in the USA: the Commingled Real Estate Fund (CREF). The first CREF was established in 1968. A CREF represents a pool of capital provided by a number of pension funds assembled to acquire property. CREFs have been established primarily for pension funds by life insurance companies, commercial banks and independent investment companies. Such funds may be open-end or closed-end.

As is the case in the UK, open-end CREF property assets are valued on a regular basis. Unlike the UK, the frequency is usually annual with independent valuers and quarterly with internal valuations. The appraisal unit value serves as the current price for both entry and withdrawal. For closed-end funds, participants acquire a share in a pool of properties that were purchased at market prices. Because closed-end funds do not admit investors after the initial amount of capital needed to undertake the investment programme is raised, frequent valuations are not needed.

In terms of diversification, such funds provide similar benefits to property companies. As the price of the units is based on valuations of the underlying properties, they should track the price evolution of the underlying assets with a low volatility due to appraisal smoothing. The main disadvantage of the units of such funds is their lack of liquidity. It might take a long time to sell the units if the market is bearish.

11.4 Unitisation

The basic idea of unitisation is to split up ownership of one property. This creates a security which carries an entitlement to some share of capital growth and income from the property in proportion to the size of the unit. This enables investors to acquire an interest in individual properties. With such vehicles an investor can choose the properties in which s/he wants to invest, which is not the case when property company shares are purchased. If an investor wants to build a diversified property portfolio, however, s/he will have to purchase units for many buildings. With property companies, one share of a well-diversified company would enable this.

When such vehicles were developed it was thought that a central trading market for units would overcome the inefficiency of the direct property market. Prices would be displayed on screens and could easily be monitored and would be determined in trading rather than by valuation. The unit cost would be small and so more potential investors could participate in the market. If there were sufficient interest in the secondary market, transaction time would be very short, the market would be much more liquid and, accordingly, transaction costs would be lower.

Several single asset vehicles were developed in the UK, but failed to become established either due to their lack of tax transparency or to market conditions at the time of their launch. The vehicles which were developed include Single Asset Property Companies (SAPCOs), Single Property Ownership Trusts (SPOTs) and Property Income Certificates (PINCs).

The SAPCO is a property company where the investment is in a single property. SPOTs are similar to property unit trusts. The income from the property is passed to the unit holders who hold a legal interest in the property. With PINCs, investors do not directly own a proportion of the property. Rather, they own securities conferring comparable benefits: an income certificate to receive a share of the rental income stream and an ordinary share in the management company set up to administer the property.

Limited partnerships represent an alternative way of investing in a property (or a portfolio of properties). In common with other forms of indirect investment, such vehicles enable investors who have capital but have no expertise and time to spend operating or administering properties to gain exposure to the property market. Limited partners are only liable to the extent of the capital they have contributed to the partnership. Limited partnerships are widely used in the USA and have more recently proved popular in the UK.[3]

11.5 Debt instruments

Rather than taking an equity position in a property or a portfolio of properties, investors can participate in the property market by lending money, for example, to a developer. From the borrower's perspective, as the acquisition cost of property generally exceeds the amount of equity capital available to individual investors, borrowed funds are necessary in order to purchase a property. Also, if the return on the investment with a 100% equity financing exceeds the after tax mortgage interest rate, the borrower may benefit from a gearing (leverage) effect but faces substantially greater risk.

A property is purchased for £4m in 1994. The annual after tax cash flows amount to £380,000. The time horizon for the investment is four years, and the sale price of the property in 1998 is equal to £4.5m (net of capital gains tax). The cash flows of the investment are, thus, as follows:

	1994	1995	1996	1997	1998
Cash flows	−4,000,000	380,000	380,000	380,000	4,880,000

The Internal Rate of Return (IRR) with a 100% equity financing is equal to 12.1%.

Now consider a mortgage loan of 70% of the purchase price (loan-to-value ratio = 70%). The mortgage interest rate is 9%, and the loan is fully amortised in 1998. The annual after tax interest payments (tax rate = 20%), are:

$$£4,000,000 \times 0.7 \times 0.09 \times (1 - 0.2) = £201,600$$

and the annual cash flows are:

$$£380,000 - £201,600 = £178,400$$

As the loan-to-value ratio is 70%, the equity input in 1994 is:

$$£4,000,000 \times 0.3 = £1,200,000$$

The 1998 cash flow is:

$$£4,500,000 + £178,400 - £2,800,000 \text{ (the mortgage repayment)}$$
$$= £1,878,400$$

The cash flows under this financing hypothesis are:

	1994	1995	1996	1997	1998
Cash flows	−1,200,000	178,400	178,400	178,400	1,878,400

The IRR on the equity amounts to 22.4%. The difference between 12.1% and 22.4% is due to the gearing effect. Again, the gearing effect will only be positive if the IRR with a 100% equity financing is greater than the after tax mortgage interest rate. If mortgage interest rates were at 16%, after tax rates would equal: $0.16 \times 0.8 = 12.8\%$ (with a tax rate of 20%). The gearing effect will be negative as 12.8% > 12.1%. The computations will be the same as above, except that the interest rate is now 16%. The cash flows would be:

	1994	1995	1996	1997	1998
Cash flows	−1,200,000	21,600	21,600	21,600	1,721,600

The IRR on the equity is 10.7%. The gearing effect is negative as debt financing diminishes the IRR (12.1% with 100% equity financing versus 10.7% with 30% equity/70% debt financing).

The greater the loan-to-value ratio, the greater the gearing effect (if IRR with 100% equity ≠ after tax interest rate). Assume that a bank would be willing to lend 90% of the value of the property at 9%. The equity input becomes:

$$£4,000,000 \times 0.1 = £400,000$$

The annual cash flows are:

$$£380,000 - [£4,000,000 \times 0.9 \times 0.09 \times (1 - 0.2)] = £120,800$$

The 1998 cash flow equals:

$$£4,500,000 + £120,800 - £3,600,000 \text{ (the mortgage repayment)}$$
$$= £1,020,800$$

The cash flows would be:

	1994	1995	1996	1997	1998
Cash flows	−400,000	120,800	120,800	120,800	1,020,800

The IRR on the equity now amounts to 46.4%.

Great caution has to be exercised with gearing as the risks increase significantly with the level of gearing. If the IRR on a 100% equity investment becomes less than the after-tax mortgage interest rate (because of higher vacancy rates and/or because of higher interest rates), high loan-to-value ratios can lead to very low (or negative) returns. As a consequence, lenders do not lend more than a given loan-to-value ratio. For commercial properties in the UK this limit is generally 60% to 70%. In the USA, it varies by property type.

For an investor, participation in the property market through lending is quite different from an equity investment in property. With conventional debt instruments, the return for the lender is not influenced by the behaviour of the property market, but is specified in the loan agreement. In this respect, property lending is very similar to fixed-income investments such as corporate bonds. With property debt instruments, the property acts as security for the debt. If the borrower defaults, the lender may foreclose. Only if the sale price of the property lies below the outstanding balance of the loan will the lender lose part of the investment.

As is the case for equity investments in property, there are direct and indirect debt instruments. Indirect debt instruments (mortgage-backed securities and mortgage REITs) mostly exist in the USA. Some general features of these two types of vehicle are presented below.

Direct debt instruments

Lending can be undertaken over the short term to finance development projects or over the long term to finance property investments. The period of development lending varies from one country to another. In the UK the period is from 2 to 5 years, and in the USA from 1 to 5 years. The traditional long-term borrowing instrument is the mortgage debenture. A firm lends money and secures its lending against the title of one or more properties.

There are several types of specification that need to be included in a mortgage agreement. These include whether it is a recourse loan or a non-recourse loan and whether it is a fixed or variable rate mortgage. Each of these is defined briefly below.

- ▶ Recourse loan: loan which has the security not only on the relevant property but also on the borrower's assets.
- ▶ Non-recourse loan: loan which only has the property as security.
- ▶ Fixed rate mortgage: the interest rate on the mortgage is fixed over a certain period.
- ▶ Variable (or floating) rate mortgage: the interest rate of the mortgage varies with the general level of interest rates and, therefore, decreases the lender's interest rate risk. However, variable rate mortgages can increase the default risk of a mortgage because the borrower may not be able to service the debt if payments increase significantly.

The typical commercial property loan is a balloon loan, where the borrower pays only interest during the life of the loan, repaying capital at maturity. This induces risk for the lender as the value of the property securing the loan may have fallen to a level at which it is not possible to refinance or sell the property and return the capital. Lenders limit their risk in two ways. First, they restrict the loan-to-value (LTV) ratio to a certain level. Second, they compute the debt coverage ratio (net operating income/debt service) and may decide upon a minimum level for that ratio. In the USA, for instance, lenders usually require a debt coverage ratio of at least 1.2. In the USA, in a balloon mortgage, payments are typically based on a 25-year to 30-year amortisation schedule, but the loan matures in 5, 7 or 10 years. It is not uncommon for commercial mortgages to have adjustable interest rates. The index on a floating rate mortgage is typically the prime rate or the LIBOR (London InterBank Offer Rate).[4]

Hybrid debt-equity instruments have been developed which enable the investor to participate in market performance. Participating (or participation) mortgages, for instance, offer a participation in the property's operating cash flows, a participation in the cash flows from the eventual sale of the property, or both. Other debt-equity instruments include joint ventures, sale–leasebacks and convertible mortgages.

A joint venture is similar to a participating mortgage in that the lender expects to receive a portion of the cash flows from operation or sale of the property as well as a scheduled mortgage payment. With a joint venture, the lender acquires an ownership interest in the property. In a sale–leaseback agreement, an institution agrees to purchase a property. The property is then leased back to the user of the property, who becomes a tenant. The tenant often has an option for the repurchase of the property. Convertible mortgages give the lender an option to convert the mortgage debt to equity after a specified period.

It is interesting to identify the big commercial lenders. In the USA, banks provide 41% of commercial funding, life companies 20%, Savings and Loans (S&Ls) 11%, pension funds 2%, REITs 1% and 'other' investors the remaining

25% (figures as of 1995, reported by Han, 1996). As of the end of 1995, the total amount of outstanding commercial debt was $981bn. Thirteen per cent of that figure was by foreign investors. Banks and insurance companies are also major players in the commercial lending market in the UK, the Netherlands, Canada and several other countries. In Switzerland, almost all of the commercial financing is provided by the banks.

Indirect debt instruments

Indirect debt instruments exist almost exclusively in the USA, where there are two main types of indirect debt instrument: the Mortgage REITs (MREITs) and the Commercial Mortgage-Backed Securities (CMBSs). As of 1995, MREITs represented approximately 1% of total commercial debt funding in the USA and CMBSs 8% (Han, 1996; Corgel et al., 1998). Mortgage REITs are REITs which invest in mortgages either directly or indirectly by purchasing mortgage-backed securities. Investments are made both in the residential and commercial sectors. The market for CMBSs is growing rapidly. As reported by Corgel et al. (1998):

'the market for CMBSs is one of the most dynamic and fastest-growing sectors in the capital markets. Although the value of outstanding commercial mortgages being used as collateral for CMBSs is only approximately 8% of total outstanding commercial debt, the size of the market is significant given CMBSs were virtually non existent prior to 1990.'

Mortgage-backed securities constitute a means of taking an indirect debt position in property. Approximately 50% of residential mortgage loans in the USA are sold into the secondary mortgage market and used as collateral for the issuance of mortgage-backed securities (MBSs). The amount of residential MBSs as of 1995 was approximately $1.8 trillion. The development of the CMBSs market has been slower than that of the residential MBS market. The rate of growth of CMBSs, however, has accelerated substantially since the beginning of the 1990s.

CMBSs are securities which are backed by a pool of commercial mortgages or a single large commercial loan. Most CMBSs take the form of a mortgage-backed bond (MBB), a mortgage pass-through security (MPT) or a collateralised mortgage obligation (CMO). With mortgage-backed bonds, a pool of mortgages is established and bonds are issued. The issuer retains ownership of the mortgages. MPTs represent an undivided ownership interest in a pool of mortgages. Cash flows are passed through to the investors. Collateralised mortgage obligations are instruments that are issued using a pool of mortgages for collateral. The issuer usually retains the ownership of the mortgage pool. Cash flows are passed through to the investors as is the case with MPTs, but the structure of the asset is more sophisticated.[5]

Contrary to what is usually the case for residential MBSs, CMBSs carry no government guarantees. Thus, borrower default constitutes the greatest risk to CMBS investors. To monitor this risk, investors are provided with ratings for CMBS issues. Commercial rating agencies include Standard & Poor's and Moody's. These securities are mostly purchased by pension funds and are considered to be illiquid investments.

11.6 Property derivatives

Derivatives are contracts whose payoffs depend on the price of some underlying asset. The value of the derivative instrument depends on the value of the underlying asset. Derivatives can take many forms such as, for instance, options, futures and forwards. Future and forward contracts are similar to options in that they specify purchase or sale of some underlying asset at some future date (the contract maturity). The key difference is that the holder of an option to buy (to sell) is not compelled to buy (to sell) and will not do so if the trade is unprofitable. A futures or forward contract, however, carries the obligation to go through with the agreed-upon transaction (see also Chapter 12). Derivatives have been created for a variety of underlying assets such as commodities, currencies and financial assets.

This example refers to a futures contract on a stock index (the FTSE 100 index), but a future on a property index would work in a similar fashion. As of 6 July 1998, the value of the future on the FTSE 100 stock index was £6053 for a maturity in September (the maturity is on the third Friday of the month). The value of the FTSE 100 stock index as of 6 July 1998 was £5990.30. Each futures contract is for 25 times the index. Thus the purchaser (long position) of a September contract can buy in September 25 times the index at $25 \times £6053 = £151,325$. The seller (short position) will agree to sell 25 times the index at that price.

The purchaser of the contract can sell it at any time on the market (settlement). If the value of the contract at that date lies above £6053 s/he will make a profit. Assume that the value of the contract is equal to £6500. The profit would amount to:

$$\text{Profit} = 25 \times (£6500 - £6053) = £11,175$$

If the value of such a contract lies below £6053 s/he will suffer a loss. If the value of the contract is £5500 for instance, the loss will amount to:

$$\text{Loss} = 25 \times (£6053 - £5500) = £13,825$$

For the seller of the contract (short position) the reverse holds, that is s/he will benefit in declining markets and benefit in rising markets. In this example, s/he will gain £13,825 if the value of the contract is £5500 and will lose £11,175 if the value of the contract is £6500. If the contract reaches maturity, the purchaser of the contract will benefit if the index at the time of maturity lies above £6053, while the seller will benefit if the index lies below that level.

In Chapter 12, currency derivatives are presented as ways to hedge against foreign currency risk. This is of importance to investors who wish to invest in international property and who wish to hedge themselves against foreign currency risk. This section examines how derivatives could be used to hedge property risk or as a means of including property in a portfolio. As such derivatives have property as the underlying asset, they are named property derivatives.

Property derivatives would enable property owners to hedge away property risk. In order for the underlying index to be as similar as possible to the properties which need to be hedged, and as advocated by Case *et al.* (1993), derivatives for each of the major geographic regions and property types would have to be created. In order to hedge their portfolio, property owners would contract to sell (that is, take a short position on) property derivatives that are closely correlated with the properties that they own.

Such derivatives could also be used for other purposes, such as to gain quick exposure to the property market, to create artificial property portfolios if there were a range of derivative contracts and for speculative reasons. Speculators use derivatives to profit from movements in future prices. If speculators believe property prices will decrease (that is, if they are bearish), they will take a short position. Alternatively, if they think prices will increase (that is, if they are bullish), they will take a long position (they will contract to buy).

In derivatives markets for commodities and financial assets, speculators use derivatives markets rather than the underlying markets for two main reasons: transactions costs are lower and the gearing is far greater. The higher gearing (leverage) means investors can make large gains (or losses) from low initial investment. For instance, a 1% variation of the underlying asset might lead to a 10% or 20% variation in the price of a futures contract.[6] Other reasons for gaining exposure to the property market through a derivatives market include low unit costs and an exposure to the market without incurring high levels of specific risk.

In the case of property, one problem is the determination of the underlying asset. As physical delivery is hardly possible, an alternative is required. This could be a cash settlement on the basis of an index of properties. One possibility would be to use a valuation-based property index. However, as noted by Shiller (1993), such indices exhibit serious drawbacks in the context of derivative instruments. They are expensive to construct if the appraisals are to be thorough. To be representative, they must be based on a large sample. In the USA, no such indices are available for a broad spectrum of residential properties. For commercial property, the indices are based on samples of properties, not randomly selected.

Derivative contracts should be settled on the basis of indices that rely on market transactions rather than the judgements of appraisers. Such indices were reviewed in Chapter 4 and include hedonic indices, repeat sales indices or hybrids of the two. Another key condition for the successful development of property derivatives is the establishment of an active secondary market: this requires both sufficient market capitalisation and investors prepared to trade actively in the market.

In 1991, a residential property futures contract that was cash settled based on the Nationwide Anglia House Price Index (a hedonic index) and a commercial futures contract which was cash settled on the Investment Property Databank (IPD) monthly index were created on the London FOX. Dealing was suspended in October 1991 after allegations of illegal trading. In the late 1990s, further attempts to introduce derivative instruments were made, with the launch of Barclays PICs and Property Index Forward, again based on IPD indices (see also Ball *et al.*, 1998).

As discussed above, property derivatives entail several benefits for property owners and investors wishing to gain exposure to the property market. Several authors (Case *et al.*, 1993; Riddiough, 1995) advocate the launch of property derivatives. Noticeably, it is argued that the development of derivatives markets for property should reduce transactions costs and diminish the inefficiency of the property market (Shiller, 1993; Case *et al.*, 1993).

11.7 Summary and conclusions

Direct property investments entail several drawbacks which include large unit costs, illiquidity, high transaction costs and the need for local market knowledge. For these reasons, investors might seek alternative ways of gaining exposure to the property market. Three broad categories of alternative ways of investing in 'property' exist: indirect equity investments, debt instruments and property derivatives.

Indirect equity investments encompass investments in property company shares, in other collective equity investment vehicles and securitisation of single assets. Property companies engage in investment and development/trading activities. The closest substitute to direct property investment should be provided by these property companies which invest in income-producing property. Such companies exist in most developed countries and indices for these markets are available.

Shares of property companies do not exhibit the disadvantages of direct property investments. For instance, such shares are more liquid and their unit value is substantially lower than that of direct investments. The price of these shares is known at all times as they trade on an exchange. In most countries there is no full tax transparency as property companies are taxed, although US REITs are an exception. The capital returns on property company shares have been shown not to track very well the returns on direct investments in several countries. In the USA, this is true despite the strong relation between indirect and direct income returns.

Other indirect equity investment vehicles include those for which there is no formal secondary market and the price of such units is based on regular valuations. Examples are Property Unit Trusts (PUTs) in the UK and Commingled Real Estate Funds (CREFs) in the USA. These vehicles are mainly designed for institutions. The basic idea of unitisation is to split up ownership of one property. Several single-asset vehicles have been developed in the UK but have failed to become established either due to their lack of tax transparency or to market conditions at the time of their launch.

Taking a debt position in a property bears a less obvious relationship with the property market. The return for the lender is not influenced by the behaviour of the property market (except for default), but is specified in the loan agreement. In this respect, such vehicles are closely related to fixed-income securities. The property acts as a security for the debt, which implies that the evolution of the property market will affect the financial position of the lender. Hybrid debt-equity instruments have also been developed which enable the investor to participate in market performance.

As is the case for equity property investments, debt positions can be taken either directly or indirectly. Indirect vehicles are common in the USA with Commercial Mortgage-Backed Securities (CMBSs) and Mortgage Real Estate

Investment Trusts (MREITs). Lending funds to a property owner can be viewed as an investment which should yield a lower return but also a lower risk than an equity investment in property over the long run.

Property derivatives would represent a way for an owner to hedge away property risk. In other words, the combination of properties and derivatives could result in near immunity against price changes. They also provide a way for investors to participate in the property market with limited funds, low transaction costs and high gearing. As such, these investments are speculative: the expected returns are high, but the associated risks are also high. Property futures were launched in London in the early 1990s, but dealing was suspended shortly after allegations of illegal trading. More recently, forward contracts have been launched.

Notes

1. For a review of the literature on price discovery, see Chau *et al.* (1998).
2. In the UK, for example, Property Unit Trusts (PUTs) are valued monthly.
3. The attractiveness of limited partnerships in the USA has depended on the tax laws. For general discussions of the effects of tax laws, see Hendershott *et al.* (1987) and Follain *et al.* (1987, 1992).
4. For more details, see Corgel *et al.* (1998), Chapter 18.
5. For details, see Brueggeman and Fisher (1997), Chapters 18 and 19.
6. With derivative instruments, the same nominal profit (loss) can be made with a much lower investment. Thus, the percentage return will be much greater with the derivative instrument. This implies that the derivative instrument is a highly geared (leveraged) instrument.

Further reading

Ball, M., Lizieri, C. and MacGregor, B. D. (1998) *The economics of commercial property markets*, Routledge, London: Chapter 12.

Barkham, R. (1995) The performance of the UK property company sector: a guide to investing in British property via the public stock market, *Real Estate Finance*, 12(1), 90–8.

Brueggeman, W. B. and Fisher, J. D. (1997) *Real estate finance and investments*, Irwin, Chicago (tenth edition): Chapters 11, 17, 18, 19 and 20.

Corgel, J. B., McIntosh, W. and Ott, S. H. (1995) Real estate investment trusts: a review of the financial economics literature, *Journal of Real Estate Literature*, 3(1), 13–43.

Corgel, J. B., Smith, H. C. and Ling, D. C. (1998) *Real estate perspectives: an introduction to real estate*, Irwin/McGraw-Hill, Boston (third edition): Chapters 5, 18 and 19.

Downs, D. H. and Hartzell, D. J. (1995) Real estate investment trusts, in *The handbook of real estate portfolio management*, Pagliari, J. L. (Ed.), Irwin, Burr Ridge (IL), 597–634: Chapter 14.

Gehr, A. K. (1995) Applications of derivative instruments, in *The handbook of real estate portfolio management*, Pagliari, J. L. (Ed.), Irwin, Burr Ridge (IL), 1112–52: Chapter 27.

Lizieri, C. and Satchell, S. (1997) Interactions between property and equity markets: an investigation of linkages in the United Kingdom 1972–1992, *The Journal of Real Estate Finance and Economics*, 15(1), 11–26.

Lizieri, C., Satchell, S., Worzala, E. and Dacco, R. (1998) Real interest regimes and real estate performance: a comparison of UK and US markets, *The Journal of Real Estate Research*, 16(3), 339–56.

Oppenheimer, P. H. (1996) Hedging REIT returns using the futures markets, *Journal of Real Estate Portfolio Management*, 2(1), 41–53.

Robinson, T. E. (1990) Real estate investment trusts, in *The real estate handbook*, Seldin, M. and Boykin, J. H. (Eds), Dow Jones-Irwin, Homewood (IL) (second edition), 740–55: Chapter 41.

Shiller, R. J. (1993) *Macro markets*, Oxford University Press, Oxford: Chapter 3.

Snyderman, M. P. (1994) Commercial and multi-family real estate investment vehicles, in *Managing real estate portfolios*, Hudson-Wilson, S. and Wurtzebach, C. H. (Eds), Irwin, Burr Ridge (IL), 50–97: Chapter 2.

The Journal of Real Estate Research (1995) Special issue on Real Estate Investment Trusts, Kuhle, J. L. (guest editor), 10(3/4).

Chapter 12

International property investment

12.1 Introduction

Chapter 9 was devoted to the optimal way of constructing a domestic property portfolio. In Chapter 10, the benefits of including property in portfolios of financial assets were examined. It was shown that property can play a substantial role in diversifying multi-asset portfolios. Chapter 11 reviewed the alternative ways of investing in property. This chapter examines whether investors should increase their range of investments further by considering international property investments. Intuitively, such a strategy is appealing as country-specific risk should be diversified away. Also, in countries where the commercial property market is of limited size, such a strategy offers the possibility for institutions with substantial capital to invest in property.

In the post-war period three main economic zones have emerged: North America (the USA and Canada), the Pacific Rim and Europe. Since before the Second World War the US economy has been the dominant world economy and remains so today. However, its relative importance has been falling, mainly as a consequence of the rise of the Far East and the Asian Pacific Rim which include countries such as Japan, South Korea, Taiwan and Singapore. During the last 15 years, economic growth in that region was double that of the other two economic zones, and consequently the Pacific Rim area is an increasingly important region for investment.

As Europe moves to integration and to a single currency, this may lead to convergence in economic and political structures. The population of Europe is now larger than that of the USA and the GDP of both areas is comparable ($8360.8bn for the European Union and $8178.8bn for the USA as of 1998, according to the OECD). The growing integration of Europe has simplified the development of a European investment strategy for non-European investors. Thus, the scope and need for diversification within Europe should gradually decrease (see Baum and Schofield, 1991).

The 'prudent man rule' (the Employment Retirement Security Act of 1972, ERISA) for US pension plans allows for more extensive international diversification and the proportion of foreign assets in US pension plans has been rapidly increasing, although it has not yet reached the level of UK or Dutch pension funds. Such investment is predominantly in the share markets.

Another explanation for the increase in international investment is the globalisation of economic activity mainly through deregulation, liberalisation and advances in information and communications technologies. Industries have developed truly international markets through mergers and acquisitions. As a consequence, the demand for overseas property by such companies is great. Although the insurance and pension industries remain primarily domestic, as they globalise, institutions with liabilities in several countries will need assets in these countries to match these liabilities.

Although measuring the extent of foreign property investment in a country or world-wide is by no means an easy task, some figures are provided by Dohrmann (1995) and Ball *et al.* (1998). The amount of foreign capital in US property, for instance, is estimated to be $34.9bn in 1990 with $15.4bn in the hands of Japanese investors.

The UK too has seen the emergence of foreign investors in the commercial property market. Most of the overseas interest has been for the Central London office market and has come mainly from German, Japanese and American investors. Foreign holdings of City of London offices are estimated at £3.6–4.5 billion by Baum and Lizieri (1998).

UK funds have a long tradition of foreign property investment. In earlier years, mainly the US property market was considered. Initially the investment was achieved through pooled funds (the commingled real estate funds, CREFs, in the USA) – for details see Chapter 11. By the 1980s, closed-end funds became more popular. These were the first stages towards direct investment in overseas property. More recently, attention has turned to a wider range of overseas markets.

Diversification is the primary purpose of international investment as individual country returns are less than perfectly positively correlated. Other risk and return factors are also important – for example, differing lease structures or tax rates across countries. For investors in some smaller countries, international property is often a necessity as their domestic property markets are too small. However, international property investment also has drawbacks caused, for instance, by the differing legal frameworks and political risks.

Managing international portfolios requires a basic understanding of equilibrium relationships that should apply to product prices, interest rates and exchange rates. These relationships form the basis of international finance (a more detailed presentation of these concepts is provided in Shapiro, 1996). This chapter begins with a review of these relationships in Section 12.2. The case for international diversification is made in Section 12.3 and the empirical evidence for shares, bonds, direct and indirect property is reviewed in Section 12.4. Section 12.5 is concerned with the ways to construct an overseas portfolio and includes a discussion of the problems associated with an international strategy. Section 12.6 contains a summary and conclusions.

If transaction costs are ignored, exchange-adjusted prices for identical tradable goods (including financial and real assets) should be the same across countries.[2] This is known as the law of one price. If this were not the case, arbitrage opportunities would exist, which means that an asset could be purchased for instance in country A where its price is low and sold in country B where its price would be higher. By acting in this manner, so-called 'arbitrageurs' would increase the asset's price in country A (the demand for that asset would be higher) and lower its price in country B (its demand would be lower). This would continue until prices were the same in both countries. In a similar way, risk-adjusted returns on assets in different markets should be equal.

Five theoretical economic relationships result from such arbitrage activities. These relationships are:

▶ purchasing power parity
▶ the Fisher effect
▶ the international Fisher effect
▶ interest rate parity
▶ forward rates as unbiased estimators of future spot rates.

All these relationships relate money supply growth, inflation, interest rates and exchange rates. The common theme of these parity conditions is the adjustment of the various rates and prices to inflation. A presentation of these relationships is provided below. A good understanding of these relations is needed to analyse the relationships among the various domestic and foreign monetary variables and their influences on asset prices. Before proceeding, however, the concept of exchange rate needs to be explained.

Exchange rates are market-clearing prices that equilibrate supplies and demands in foreign exchange markets.[3] The exchange rate indicates how much of one currency is needed to buy or sell one unit of another currency. An exchange rate of £1 = $1.5 indicates that £1 will buy $1.5. If the exchange rate becomes £1 = $1.6, the pound is said to have appreciated and the dollar to have depreciated.[4]

Home currency appreciation relative to foreign currencies is the international equivalent of domestic inflation. In the same way that the price of goods in one year cannot be compared with the price of goods in another year without adjusting for inflation, exchange rate changes and interest rate changes may indicate nothing more than the reality that countries have different inflation rates.

Purchasing power parity

Purchasing power parity states that exchange-adjusted price levels should be identical world wide. In other words, a pound sterling should have the same purchasing power around the world, because free trade should equalise the price of any tradable good in all countries. This is only true, however, if transportation costs, restrictions to free trade and product differentiation are not considered. The relative version of purchasing power parity states that the exchange rate between the home currency and any foreign currency will adjust to reflect changes in the price levels of the two countries. This relationship can formally be written as:

$$\frac{e_t}{e_0} = \frac{(1 + i_h)^t}{(1 + i_f)^t}$$

◀Equation 12.1

where: e_t and e_0 is the exchange rate (that is the value in the home currency of one unit of foreign currency) at time t and 0, respectively
i_h is the inflation rate in the home country
i_f the inflation rate in the foreign country.

If the UK and the USA are running annual inflation rates of 3% and 2%, respectively, and the exchange rate is £1 = $1.5 (if the home currency is the pound sterling, the exchange rate is $1 = £0.667), then the value of the dollar in a year (e_1) should be:

$$e_1 = 0.667\left(\frac{1.03}{1.02}\right) = 0.673$$

The one-period equation is often represented by the following approximation:

$$\frac{e_1 - e_0}{e_0} = i_h - i_f$$

◀Equation 12.2

and the exchange rate change during a period should equal the inflation differential for the same time period. Exchange rate movements should thus just cancel out changes in the foreign price level relative to the domestic price level. Purchasing power parity says that currencies of countries with high inflation rates should devalue relative to currencies of countries with lower rates of inflation.

In reality, exchange-adjusted prices will differ across countries because of transportation costs and barriers to trade. Empirical studies have demonstrated, however, a clear relationship between relative inflation rates and changes in exchange rates. Shapiro (1996), for instance, compares the relative change in the purchasing power of 22 currencies with the relative change in the exchange rates of these currencies for the period 1982–8. Countries with high inflation rates (such as Brazil, Israel or Peru) saw sharp erosion in their foreign exchange values. Similarly, countries with low inflation rates (such as Japan, Switzerland or West Germany) experienced strong currencies. These empirical studies show that purchasing power parity holds up well in the long run, but not as well over shorter time periods. Short-run deviations from purchasing power parity occur, but currencies have a tendency to move towards the rates predicted by purchasing power parity.[5] Also, price indices heavily weighted with non-traded goods and services are misleading. For instance, changes in relative prices of hotel rooms around the world should have little impact on exchange rates. If you are staying in London you will not be able to substitute a hotel room there for a room in a city where prices are lower.

The Fisher effect

The Fisher effect states that the nominal interest rate r comprises two elements: a real required rate of return and a premium for expected inflation (see also Chapters 2 and 5)[6]:

$$(1 + r) = (1 + a) \times (1 + i)$$

This equation can be rewritten as:

$$r = a + i + ai \qquad \qquad \blacktriangleleft \text{Equation 12.3}$$

where: r is the nominal interest rate
   a is the real interest rate
   i is the expected inflation rate.

As the cross-product of a and i is usually small it is discarded and the nominal interest rate is approximated by the sum of the real interest rate and expected inflation. If the real required interest rate is 5% and the expected inflation rate 2%, the cross-product would amount to $0.05 \times 0.02 = 0.001 = 0.1\%$ and would be neglected, giving a nominal interest rate of 7%.

The generalised Fisher effect asserts that real interest rates should be equal in all countries otherwise arbitrage opportunities would exist. If expected real returns were higher in country A than in country B, for instance, capital would flow from country B to country A until real interest rates equalised. In equilibrium, the nominal interest rate differential between the home and the foreign country will approximately equal the expected inflation rate differential:

$$\frac{1 + r_h}{1 + r_f} = \frac{1 + i_h}{1 + i_f} \qquad \qquad \blacktriangleleft \text{Equation 12.4}$$

where r_h and r_f are the nominal interest rates in the home and foreign country, respectively.

If r_f and i_f are relatively small, this equation can be approximated by

$$r_h - r_f = i_h - i_f \qquad \qquad \blacktriangleleft \text{Equation 12.5}$$

Currencies of countries with high inflation rates should bear higher interest rates than currencies with lower rates of inflation. If country A's expected inflation is 30% and country B's expected inflation is 2%, the nominal interest rate in country A should be 28% higher than in country B. Empirical studies for the period 1982–8 have shown that this relationship holds in general. Nominal interest rates in Mexico and Peru, for instance, are substantially higher than in Japan or Switzerland. Parity might not hold, however, if investors preferred domestic assets in order to avoid currency risk, even if the expected real return on foreign assets were higher. Real interest rates can also differ because of varying degrees of political risk. Real interest rates in some developing countries, for example, can exceed those in developed countries without offering attractive arbitrage opportunities to foreign investors. Similarly, the comparison of real interest rates must be undertaken for assets of similar risk. Real share returns, for instance, cannot be compared with real interest rates on Government bonds.

The international Fisher effect

The international Fisher effect arises from combining the purchasing power parity and Fisher effect relationships and states that there is a relationship between nominal interest rates and exchange rates:

$$\frac{(1 + r_h)'}{(1 + r_f)'} = \frac{\bar{e}_t}{e_0}$$

◄Equation 12.6

where $\bar{e}_t$ is the expected exchange rate in period t. If r_f is relatively small, this equation can be approximated as follows for one period:

$$r_h - r_f = \frac{\bar{e}_1 - e_0}{e_0}$$

◄Equation 12.7

Currencies with low interest rates are, thus, expected to appreciate relative to currencies with high interest rates, and the nominal interest rate differential should be an unbiased estimator of the future change in the spot exchange rate. This relationship has been shown to hold in the long run with currencies of countries such as Israel (where interest rates are high) depreciating and currencies of countries such as Switzerland or Japan (where interest rates are low) appreciating. As mentioned for the Fisher effect, this relationship only holds if investors do not require a risk premium to hold foreign assets.

Interest rate parity

The parity relationships which have been reviewed so far only consider the spot exchange rate. In fact, there is also a forward exchange market. Whereas in the spot market currencies are traded for immediate delivery, in the forward market contracts are made to buy or sell currencies for future delivery. A major determinant of the spread between forward and spot rates is the interest rate differential between countries. According to interest rate parity, the currency of a country with a lower interest rate should be at a forward premium in terms of the currency of the country with the higher rate. In other words, the interest differential between the home country and a foreign country should be approximately equal to the forward differential. This can be written as follows:

$$r_h - r_f = \frac{f_1 - e_0}{e_0}$$

◄Equation 12.8

where f_1 is the end-of-period forward exchange rate, that is the exchange rate which can be used in period 1 if such a contract is signed. This rate should not be mixed up with the spot exchange rate which will prevail in period 1 (which is denoted by e_1).

Interest rate parity ensures that the return on a hedged, or covered, foreign investment will just equal the domestic interest rate on investments of identical risk.[7] If this were not the case, arbitrage opportunities would exist and there would be an incentive to move money from one market to the other. High interest rates on a currency are offset by forward discounts and low interest rates are offset by forward premiums. If the UK interest rate is 6% and the US interest rate 4%, for instance, there should be a forward discount on the exchange rate of the pound sterling (that is, the forward exchange rate should be lower than the spot exchange rate). Tests of interest rate parity have been widely undertaken and document the fact that deviations from parity tend to be small and short-lived. In fact, in the Eurocurrency markets, the forward rate is calculated from the interest differential between currencies.

The relationship between the forward rate and the future spot rate

The last relationship which needs to be examined is that between the forward exchange rate and the future spot rate. Equilibrium on the forward exchange market will be reached when the forward differential equals the expected change in the exchange rate. At this point, there is no longer any incentive to buy or sell the currency forward. If this is true, the forward exchange rate should constitute an unbiased estimator of the future spot exchange rate. Several studies have examined this assertion and, in general, empirical evidence suggests that this is indeed the case.

12.3 The case for international diversification

Individual domestic assets within an asset class tend to move up and down together because they are similarly affected by domestic conditions such as movements in interest rates or national growth. This creates a strong positive correlation within the asset classes. As discussed in Chapter 10, the aim of inter-asset diversification is to take advantage of the low correlation across asset classes and to diversify away some of the specific risk of the asset classes. Even when such a strategy is undertaken, there remains some risk which is specific to the country and which could be diversified away through international diversification. In this section, the arguments for considering international assets are examined, while the empirical evidence on international diversification for shares and bonds, and for property is presented in Section 12.4.

Risk diversification

By expanding the universe of assets available, international diversification should make it possible to achieve a better risk-return trade-off than by investing solely in domestic assets. In other words, expanding the universe of assets available for investment should lead to higher returns for the same level of risk or less risk for the same level of expected return. This relation follows from the basic rule of portfolio diversification: the broader the diversification, the lower the risk (see Chapter 7). Through international diversification, risk that is systematic in the context of the home economy may be unsystematic in the context of the global economy.

Going international implies that currency movements have to be taken into account. As far as the return of an international investment strategy is concerned, it can be calculated as follows:

$$1 + R_h = (1 + R_f)(1 + e)$$
◄Equation 12.9

where: R_h is the home currency return
R_f the foreign currency return
e the exchange rate of the foreign currency in terms of the home currency.

Ignoring the cross-product $R_f e$, Equation 12.9 can be approximated by the following equation:

$$R_h = R_f + e$$
◄Equation 12.10

So the home currency rate of return is approximately equal to the sum of the foreign currency return plus the change in the home currency value of the foreign currency.

A UK investor buys a US share whose price is $200. The exchange rate is £1 = $1.5, so the purchase price of that share in sterling is 200/1.5 = £133.33. Assume that the market value of that share a year later is $220, yielding a 10% total return. If the exchange rate is unchanged at the end of the year, the total return in sterling on that share is 10%. If the exchange rate has changed, then the return in sterling will be higher or lower than the return in dollar depending on whether the pound has depreciated (or the dollar appreciated) or appreciated (or the dollar depreciated).

If the dollar has appreciated by 10% (that is, the exchange rate at the end of the year is £1 = $1.364), the total return in sterling will be: 10% + 10% = 20%. This is an approximation of the true return in sterling which can be computed as follows:

Market value of the share at the end of the year (in £)	$220/1.364 = £161.29
Purchase price of the share (in £)	$200/1.50 = £133.33
Total return in sterling	$R = \dfrac{161.29}{133.33} - 1 = 21.0\%$

The difference between 20% and 21% corresponds to the cross-product 10% × 10% = 1%.

If the dollar has depreciated by 10% (that is, the exchange rate at the end of the year is £1 = $1.667), the total return in sterling will be: 10% – 10% = 0% and the dollar return would be totally offset by the currency depreciation. Again, this is an approximation of the true return in sterling which can be computed as follows:

Market value of the share at the end of the year (in £)	$220/1.667 = £131.97
Purchase price of the share (in £)	$200/1.50 = £133.33
Total return in sterling	$R = \dfrac{131.97}{133.33} - 1 = -1.0\%$

The difference of 1% is again due to the fact that in the approximation the cross-product is discarded.

If the various parity conditions held, the real return would be the same across countries for assets with a similar level of risk. In the short run these parity conditions usually do not hold and exchange rate movements imply an additional source of risk: currency risk. However, this additional source of risk does not imply that international diversification is unattractive as market risks and currency risks are not additive. They would only be additive if the two were perfectly correlated. In fact, there is only a weak, and sometimes negative, correlation between currency and market movements. The relation between the total risk and the market and currency components of risk is as follows:

$$\sigma^2 = \sigma_f^2 + \sigma_e^2 + 2\rho\sigma_f\sigma_g \qquad \qquad \blacktriangleleft \textbf{Equation 12.11}$$

where: σ the total risk in the currency of the investor (base currency)
σ_f is the market risk in the domestic currency
σ_e the exchange rate volatility
ρ the correlation coefficient between the two risks.

Since $\rho < 1$, $\sigma \leq \sigma_f + \sigma_g$ highlighting the fact that the two risks are not additive. This is exactly the same principle as that for a two-asset portfolio (see Chapters 3 and 7).

Assume an asset in France with a standard deviation of 23% (in French francs). The standard deviation of the French franc/pound sterling exchange rate is 11%. The correlation between the French franc return and the rate of change in the exchange rate is 0.3. It is easy to show that the total risk for a UK-based investor ($\sigma_£$) is substantially less than 23% + 11% = 34%:

$$\sigma_£ = \sqrt{0.23^2 + 0.11^2 + 2 \times 0.3 \times 0.23 \times 0.11} = 28.32\%$$

In this case, currency risk increases the risk. It is possible that currency risk could lower the risk of investing overseas if there were a sufficiently large negative correlation between the rate of exchange rate change and the foreign currency return.

To ascertain whether international portfolio diversification should be undertaken, the risk reduction resulting from the less than perfect correlation between assets across countries should be balanced with the additional risk (if any) due to currency movements. In most cases, this comparison is clearly in favour of international portfolio diversification. However, the above example deals with one asset only in one foreign country, whereas an investor is not likely to consider only one foreign currency and part of currency risk will get diversified away by the mix of currencies represented in the portfolio.

If the overseas investment is undertaken in order to match liabilities in these overseas markets, currency risk does not constitute a problem. Also, an investor may want to be exposed to currency risk if s/he believes that the foreign currencies will appreciate. If s/he does not want to be exposed to currency risk, the investment may be hedged for major currencies by forward or futures currency contracts, currency options, back-to-back loans, currency swaps or even borrowing foreign currency to finance the investment (see also Chapter 11 and Ball *et al.*, 1998).

A forward contract is the short term (usually up to a year) negotiated right and obligation to take delivery of a sum of foreign currency at a specified price (in domestic currency) on a specified date. Future contracts operate in a similar fashion to forwards, except that they are traded and regulated on exchanges, carry a brokerage fee and there are daily transfers of cash to account for the changing settlement value. Futures tend to have standardised amounts and delivery dates.

Options give the right, but not the obligation to buy (call option) or to sell (put option) a financial instrument at a particular price on a particular day or during some period of time. As with forwards, options can be used to lock in capital value to the domestic currency at a particular exchange rate. The advantage of options is that they need not be exercised should the exchange rate shift in the investor's favour. This advantage has a cost – a premium has to be paid in order to purchase an option.

Back-to-back loans involve firms in two countries agreeing to borrow capital for the other firm in their home markets. The foreign firm makes interest payments and repays the loan on maturity. Currency swaps operate in a similar way to back-to-back loans, although they are financial contracts generally handled through an intermediate dealer. The investor agrees to swap principal and payments. Thus, the initial principal and fixed cash flows are locked into the domestic currency. Initially, the two parties exchange equivalent amounts of two different currencies so that each has the desired quantity of foreign currency. They then make periodic interest payments to each other during the life of the contract. These payments must be made in the currency borrowed and reflect the level of interest rates in the home country of this currency. Finally, the two parties complete the swap by re-exchanging the principal sums of cash originally borrowed (for more details see Worzala, 1995; Worzala et al., 1997; Ziobrowski et al., 1997).

As mentioned by Ball et al. (1998), most of the available hedging instruments have been designed for use with short time horizon assets such as shares, bonds and cash. As a result, while they may be well suited to hedging indirect property investments, they may be less appropriate for direct ownership of foreign property. Given the high transaction costs and the illiquidity of property investments, directly held property tends to have longer holding periods than more liquid assets. The long holding period adds uncertainty to the investment, since the final sale price is more difficult to predict, making it difficult to hedge the capital appreciation. Moreover, the cost of hedging is often high and that cost should be weighted against the risk it eliminates.

Other advantages of going international

Risk reduction is the most established and frequently invoked argument in favour of international investment. However, it is not the sole motive for international investment. In most cases, international diversification lowers risk by eliminating non-systematic volatility without sacrificing expected return. In fact, expected returns in foreign markets are often higher than on the domestic market. Another advantage of international investment is offered by the mere size of foreign markets as compared to the size of some domestic markets. Given the limited size of the Dutch property market, for instance, Dutch pension funds have looked at investment opportunities outside their country, in particular in the USA.

As companies become international and employ people in a wide range of countries, the company pension plans create overseas liabilities. A growing number of pension funds are feeling the need to match these liabilities with appropriate assets. Insurance companies too are becoming global players with liabilities in several different currencies and countries. Property may represent a good asset to match these liabilities.

12.4 The empirical evidence

The empirical evidence for international diversification is examined first for shares and bonds, then for direct property and finally for indirect property.

Shares and bonds

As noted by Solnik (1993):

> 'the degree of independence of a stock market is directly linked to the independence of a nation's economy and government policies. To some extent, common world factors affect expected cash flows of all firms, and therefore, their stock prices. However, purely national or regional factors seem to play an important role in asset prices, leading to sizeable differences in the degrees of independence between markets. It is clear that constraints and regulations imposed by national governments, technological specialisation, independent fiscal and monetary policies, and cultural and sociological differences all contribute to the degree of a capital market's independence.'

Empirical studies have shown that the correlation between both share markets and bond markets in the various countries varies over time. However, the correlations are always far from unity. The correlation between UK and Japanese shares, for instance, was 0.34 over the period 1971–89, suggesting diversification opportunities (see Solnik, 1993). As expected, some markets are more integrated than others. The correlation between US and Canadian share returns, for instance, was 0.70 over the same period. Similarly, shares of countries of the Deutschmark block (Germany, Switzerland, the Netherlands and Belgium) are correlated.

The share markets in some countries are truly international, reflecting, for instance, the fact that companies quoted on these markets have to invest internationally due to the limited size of the domestic market. The Dutch share market, for example, exhibited a correlation of 0.75 with an index of world shares. Most of the Dutch market capitalisation is accounted for by international firms such as Royal Dutch-Shell, Unilever, Philips and AKZO.

Research on international share diversification has also shown that national share markets have wide differences in returns and risk and that emerging markets (that is, all of South and Central America, all of the Far East with the exception of Japan, Australia and New Zealand, all of Africa, parts of Southern Europe, as well as Eastern Europe and countries from the former Soviet Union) have had higher risk and return than the developed markets.

The standard deviation of an internationally diversified share portfolio is as little as 11.7% of that of individual securities. An internationally diversified portfolio is less than half as risky as a fully diversified US portfolio. Moreover, from 1961 to 1990, the compound annual return for non-US shares was 13.7% compared with 9.6% for the US market. The obvious conclusion is that international diversification pushes out the efficient frontier (see Chapter 7) allowing investors simultaneously to reduce their risk and increase their expected return. Shapiro (1996) calculates the risk and return of a portfolio containing 50% of US shares and 50% of non-US shares. The risk of the internationally diversified portfolio is considerably below the risk of the US portfolio and the expected return is much higher as well.

Solnik (1993) reports that adding foreign shares to a purely domestic portfolio reduces risk much faster than the risk reduction which is obtained through domestic diversification: with as few as 40 securities equally spread among the major share markets, the risk for a US investor is less than half that of a purely domestic portfolio of comparable size.

It is often assumed that as the underlying economies become more closely integrated, national capital markets will become more highly correlated. The correlation coefficients between markets are usually higher than in the 1970s. However, contrary to intuition, these correlation coefficients have fallen in recent years. Recent research has also shown that global diversification is unfortunately of limited value when markets are at their most volatile and, therefore, when investors most seek safety. However, benefits clearly exist for investors who invest for the long term.

Similar conclusions may be reached for bonds. Shapiro (1996) reports results of a study by Barnett and Rosenberg (1983) for a US investor over the period 1973–83. The portfolio's return rises and the volatility of the portfolio falls as foreign bonds are included. By investing up to 60% of their funds in foreign bonds, US investors could have raised their return substantially while not increasing risk above the level associated with holding only US bonds. As could be expected, an international share and bond diversification strategy results, for given return levels, in substantially less risk than international share diversification alone (see Solnik and Noetzlin, 1982).

Unhedged direct property

Investors who want to implement an international property diversification strategy have two possibilities: they can invest directly, by buying actual buildings, or they can invest indirectly by purchasing shares of listed property companies (see Chapter 11). Case *et al.* (1997) provide some empirical evidence on direct international property diversification. They use total returns (income and capital) on industrial, office and retail property in 21 cities in 21 different countries for the period 1986–94. Their capital appreciation figures are not based on valuations, but upon changes in capitalised asking rents (rather than effective rents which were not available).

Their return series suggest that returns to commercial property tend to move together (although not perfectly) across property types within each country. The return and risk parameters vary substantially across countries. The average total return and standard deviation on Taipei office property were 28.8% and 28.0%, respectively, while these figures were 8.7% and 9.8%, respectively, in Frankfurt.

Mean-variance analysis is performed by Case et al. (1997) for a US investor. Their results suggest that international property diversification within the three types of commercial property would have been beneficial as better risk-return trade-offs can usually be obtained. The only exceptions to this are Lisbon for industrial markets, Hong Kong and Taipei for office markets and Lisbon and Taipei for retail markets. These countries lie on the respective efficient frontiers suggesting that, for certain levels of risk, no diversification at all is required as the maximum return for that level of risk can be achieved by investing solely in property of that one city. Two issues, however, need to be mentioned. First, these results are time specific and may differ if another period were considered. Second, the limited size of some markets may make it impossible to implement the theoretical allocations given by mean-variance efficiency analysis.

Further analysis is undertaken by these authors to investigate whether groups of cities can be identified that behave differently across groups and similarly within groups (the procedure which enables such an identification of groups is known as cluster analysis).[8] For industrial and retail property, two clusters are identified, whereas for office property, six clusters are found. Efficient frontiers are constructed based on the cluster results and international diversification is again found to be beneficial. Cluster analysis is useful in devising portfolio strategies. Cities (markets) can be identified which move in a similar fashion and across which diversification may not prove useful. Alternatively, markets with a high level of dissimilarity can be identified across which diversification should prove particularly useful. The identification of such patterns permits more cost-effective portfolio strategies to be pursued.

The analysis by Case et al. (1997) can be criticised in two ways. First, only one city in each of the countries is considered, meaning that risk diversification within a country is ignored. However, the largest city in each of the countries is usually considered which would probably match the aspirations of most institutions. Second, total returns are not computed on the basis of actual prices but of capitalised asking rents. This method of estimating values implicitly assumes that new asking rents are good proxies for expected future rents; the reasonableness of this assumption will vary in time.

Quan and Titman (1997) also use data for direct property investments in 17 countries for the period 1988–94. The estimates of capital and income values are based on the consensus of opinions of a group of active market participants about the price and rents paid for prime commercial properties based on market transactions in each market. This group is polled about the prevailing price and lease terms for prime commercial property. This method should yield quite similar results as appraisal-based indices (this type of property index is discussed in Chapter 4).

Quan and Titman compute the average annual return both in the domestic currency and in US dollars and the correlation coefficients across countries but only in local currencies. They also compare the correlation coefficients between US property and property in the other countries and between US share returns

Table 12.1▶ Vacancy rates in office, industrial and retail property across selected cities

City	Office	Industrial	Retail
London	11%	5%	3%
Edinburgh	9%	8%	2%
Berlin	7%	5%	1%
Geneva	3%	9%	4%
Madrid	8%	15%	2%
Moscow	2%	NA	2%
New York	18%	8%	20%
San Francisco	9%	9%	2%
Vancouver	9%	4%	3%
Sydney	14%	7%	5%
Melbourne	22%	5%	12%
Singapore	6%	NA	12%

Note: NA = not available.
Source: ICPA – The Commercial Network (1996).

and non-US share returns. With the exception of Japan, Indonesia and the UK, share returns are generally correlated with those of the US; the property correlations show much more variability than those of share returns. The capital return on Hong Kong, Indonesia, Malaysia and Singapore property, for instance, is strongly negatively correlated with US property. The income return for New Zealand, Spain and UK property is also strongly negatively correlated with US property. These results suggest significant benefits from international property portfolio strategies (see also Newell and Webb, 1996).

The correlation coefficients across international property markets are less than unity because underlying economic and political factors vary across countries. One way of looking at this is to examine the vacancy rates across countries. Table 12.1 reports vacancy rates for office, industrial and retail property in London, Edinburgh, Berlin, Geneva, Madrid, Moscow, New York, San Francisco, Vancouver, Sydney, Melbourne and Singapore. Two other important underlying factors which vary across countries are the unemployment rate and the growth rate in GDP (see Worzala and Bernasek, 1996, for unemployment rates and GDP growth rates across selected countries).

The benefits of international diversification have been challenged by some authors. Ziobrowski and Curcio (1991), for instance, found no substantial benefits for UK and Japanese investors to purchase US property for the period 1973–87. The volatility in exchange rates created so much risk in US property returns that the additional risk totally overwhelmed any possibility of increased portfolio efficiency for these investors. The same analysis was performed by Ziobrowski *et al.* (1996) but taking into account taxes. These authors concluded that there were no significant after-tax benefits for foreign investors from investment in US property.

Caution should, however, be exercised when interpreting these results as only one foreign country (the USA) is considered for UK and Japanese investors. To be effective in reducing risk, an international strategy should involve several overseas markets. Thus, full diversification benefits are not achieved with such a strategy. Moreover, only one currency is considered. If several currencies were considered, currency risk could be partly diversified away. Thus, the study by Ziobrowski and Curcio most probably understates the risk reduction benefits from international diversification and overstates currency risk.

Hedged direct property

Several studies have investigated the benefits from hedging strategies. Worzala (1995) shows that when forward contracts are used for a US investor purchasing UK property, the investor would have obtained superior returns by not attempting to hedge when the transaction costs and the roll-over costs of the three-month forward contracts are taken into account.[9] Ziobrowski and Ziobrowski (1993) show that, as a long-term hedge, currency options are not reliable as their effect depends in large part on the direction and magnitude of the swings in exchange rates.

Worzala et al. (1997) examine the potential of a currency swap hedging strategy for a US investor purchasing a UK property investment. To account for uncertainty of exchange rates, Worzala et al. run a series of simulations with different currency scenarios. They show that hedging results in superior risk-adjusted returns as the volatility due to currency movements is eliminated. Ziobrowski et al. (1997) show that hedging US property investments with currency swaps suppresses most of the risk induced by currency instability.

These empirical studies seem to indicate that currency swaps are best suited in order to hedge overseas property investments. Hedging strategies are costly, however, and investors may prefer to expose their portfolio to currency risk and take advantage of the appreciation in foreign currencies. The studies also generally suggest that, although currency risk can be – at least partially – hedged, international diversification is not attractive in a Modern Portfolio Theory (MPT) framework (for a discussion of MPT, see Chapter 7). These studies, however, have only investigated the case of foreign property in one overseas country (for instance, the benefits of US property for a UK investor). If a full-blown diversification strategy is carried out, the risk reduction benefits should be substantially greater (as illustrated in the studies by Case et al., 1997, and Quan and Titman, 1997).

Moreover, hedging strategies may not be needed as currency risk should be diversified when several currencies are considered. Further research is needed in order to measure the exposure to currency risk and to compare that form of risk with the risk reduction benefits from an international property portfolio diversification in several overseas markets. Attention should also be paid to investigate whether hedging is profitable in such a context.

Property shares

Several studies have investigated the usefulness of constructing portfolios containing international property shares. Hartzell et al. (1997) compute the correlation coefficients between UK, US, European, Japanese, Asian and World property stock returns and common stock returns. In every instance, except between Asia and the World, the correlation between property markets is less than the correlation between the same continents' overall share markets. In many cases, the difference is significant. For instance, the correlation between UK and US share market returns is 0.60, whereas that correlation is only 0.38 for property stocks. The correlation between US and European stock returns is 0.54, but only 0.25 for property shares. Thus, there appear to be greater diversification opportunities for international property stocks than for common stocks.

As discussed in more detail in Chapter 11, Global Property Research (GPR) produces property share indices for 27 countries. These indices include all publicly listed companies with market capitalisation exceeding US$50m and with more than 75% of their revenues coming from equity real estate investment portfolios. Development and construction companies are, therefore, excluded. Eichholtz (1996) uses the GPR indices for France, the Netherlands, the UK, Sweden, Hong Kong, Japan, Singapore, Canada and the USA for the period January 1985 to August 1994 and compares these with stocks and bonds indices. All returns are expressed in local currencies and, therefore, reflect the position of an investor who is fully hedged against currency risk. However, the cost of hedging is not taken into account.

Eichholtz finds strong evidence that international property share returns are correlated less strongly than international common stock and bond returns. The efficient frontier of international property shares is constructed and compared to domestic property share strategies. Substantial benefits from international diversification are reported. For a given expected return, standard deviations for international property shares were more than 1% lower than those for the USA and France, more than 3% lower than for the UK and more than 5% lower than for Japan.

International efficient frontiers for shares and bonds are also constructed by Eichholtz (1996) and compared to international property share diversification. Despite the fact that the standard deviations of the national property share indices are, on average, somewhat higher than common stocks' standard deviations, the minimum variance property share portfolio has a more than 1% smaller monthly standard deviation than the common stock minimum variance portfolio. Thus, investments in property shares are relatively risky for portfolios limited to only one country, but international diversification is very effective in making international property share portfolios safe. Only an international bond portfolio presents lower risk, primarily due to the low standard deviations of bonds in general.[10]

Gordon et al. (1998) go one step further than to construct efficient frontiers for each of the three asset classes (shares, bonds and property). They evaluate international property shares within a framework of a mixed-asset portfolio containing US shares, US corporate bonds, US property shares and international common stocks for the period 1984–96. Consistent with previous studies, correlation coefficients between the various US assets and world property shares are found to be much lower than the correlation coefficients between the US assets and international common stocks.

The efficient frontier which contains US shares, bonds and REITs is dominated by the frontier which contains US and international stocks. This would suggest that international stock diversification would be more effective than domestic inter-asset diversification through REIT stocks. However, and as discussed in Chapter 10, REITs and US stocks are closely related and REITs are usually viewed as being a poor indicator of the direct property market. The issue as to whether international share diversification is more effective than domestic diversification through the direct property market remains unanswered. Gordon *et al.* (1998) also construct a third efficient frontier which contains domestic shares, international shares and international property shares. This latter frontier dominates the all-equity frontier making the point that international property share diversification is very effective.

Liu and Mei (1998) find similar results. They show that investing in international property-related securities provides additional diversification benefits over and above those associated with international stocks. These benefits are more pronounced at lower risk-return levels of the optimal portfolio and are present regardless of whether currency risks are hedged.

12.5 Constructing an international property portfolio

Objectives, structure and strategy

The basic principles of portfolio construction which are discussed in Chapter 9 hold whether it is a domestic portfolio or an international portfolio. Portfolio construction should proceed in the following stages:

1. Set out the objectives of the overseas investment (in terms of return and risk, what benchmark will be used, the timescale of the investment and whether the international exposure should be gained through a direct or indirect investment).
2. Decide an appropriate structure to meet the objectives based on forecasts.
3. Implement the strategy.

An important first consideration is whether the international investment is a small part only of a domestic portfolio, whether the investment is truly international, including exposure to the domestic market, or whether the investment is a stand-alone international portfolio. Each of these serves a different function: only the latter two imply a full exposure to international property markets.

If the international property investment represents a small fraction only of the whole portfolio, the international part could have a substantial amount of risk without much effect on the overall portfolio risk. This would mean that less attention has to be paid to specific risk and that there would be less need to diversify by holding a larger number of buildings. It would be possible to acquire a few buildings in each of the countries or group of countries selected. Such an international property portfolio would still have a large amount of specific risk but when included with the domestic portfolio, the impact would be marginal.

If the international property investment represents a larger fraction (or the whole) of the overall portfolio, diversification of property specific risk is more of an issue. Objectives in terms of risk and return have to be set. Also, agreement has to be found as to what benchmark should be used in order to measure the fund manager's ability for timing and selection of investments. This is no straightforward task as indices for direct property (usually appraisal-based indexes, but sometimes also based on the hedonic method) only exist in some countries. Indices of property company shares, however, exist for several countries now (see Chapters 4 and 11). One objective of international diversification might be, for instance, to outperform similar international funds. In this case, the appropriate competitors need to be identified.

An important objective of an international property investment strategy is whether the investment should be made through the purchase of buildings or indirectly through the purchase of property company shares. As noted by Eichholtz *et al.* (1997), before the late 1980s international investors were forced to choose direct investment since the size of the global indirect property market was too small to absorb substantial amounts of capital. Since then, a global market for property shares has developed which currently has a total market value of approximately US$350bn.

The appropriate structure of the international investment should be based on the objectives of the investment and would require diversification across countries or zones. However, as pointed out by Hartzell *et al.* (1993), who used employment data for the European regions, regionally diverse investments are not necessarily economically diverse as the economic base of regions can be very similar. For an international strategy to be most effective, a truly economically, as opposed to geographically, diversified property portfolio should be constructed.

Forecasts of the economies and markets, at least in broad terms, are needed. If direct investment is decided, the pricing framework (see Chapter 5) can be applied to the other countries (or other zones). Income growth can be based on economic growth (growth of GDP for instance) and the risk premium and depreciation can be considered relative to domestic figures.

Selecting the buildings requires local expertise and so local agents acting on specific instructions are needed. These agents should search buildings which meet the investor's criteria. Joint ventures using local capital and expertise also represent an attractive way of investing in direct international property markets. A joint venture with another domestic institution is also possible. Less equity is needed and so a broader diversification can be undertaken by investing a smaller fraction in more buildings.

Worzala (1994) surveyed predominantly British, Dutch and German investors who were actively considering international investments or had been active in the past. This survey shows that these investors were more inclined to choose direct investment rather than indirect investment. The preferred strategies consisted of joint ventures with other institutional investors and 100% equity investments.

Eichholtz *et al.* (1997) analyse whether a direct or indirect international property investment strategy is more profitable. The performance of a direct investment is obtained by studying property companies which have the bulk of their property assets outside their own country. These companies own an international direct property portfolio and Eichholtz *et al.* argue that the returns on such companies should reflect the performance of a direct strategy. The performance of indirect property portfolios is proxied (or approximated) by performance indices of locally operating property companies.

The Eichholtz et al. (1997) results suggest that direct investment in foreign property gives a worse risk-adjusted return than indirect international property investment. They reckon that international investment in property should be done indirectly through locally operating property companies. They conclude that an optimal property investment strategy is to invest directly at home and indirectly abroad. However, this conclusion relies on the proxy used by Eichholtz et al. for direct property. Much more research is needed on this important issue before a definite conclusion can be reached.

Another important issue is whether property returns are driven by continental factors. Strong continental factors would imply that optimal diversification can only be achieved by investing inter-continentally. Furthermore, the existence of strong continental factors would imply that international investors can acquire a near-optimal country allocation by just selecting a country from each continent. The only research which has been undertaken in this area has made use of data for property shares.

Eichholtz et al. (1993) use principal component analysis to construct groups of countries based on the co-movements of their property returns. They find evidence for continental effects, especially for Europe and North America. Property returns in the UK and Japan do not show a clear relationship to any continental group.

Eichholtz, Huisman, Koedijk and Schuin (1998) use multi-factor models and find clear evidence for the existence of continental factors in property returns. Europe especially has a very strong continental factor which reflects the growing integration of European markets. In North America, the continental factor is less strong, but significant and stable. In the property markets of the Asia/Pacific region no continental factor is found. Only Singapore's property market appears to be integrated with its neighbours.

Problems associated with an international strategy

International investment does not carry only advantages; several problems are also associated with an international property strategy. Solnik (1993) mentions that barriers to foreign investment exist, but that they are usually exaggerated. These barriers are:

1. *Familiarity with foreign markets*: Investors are often unfamiliar with foreign cultures and markets. They feel uneasy with different languages, different time zones and so on.
2. *Regulations*: In many countries regulations limit foreign investment. This was until recently the case, for instance, in Switzerland, where property could not be purchased by foreigners who did not live in the country. Institutions in some countries are also often constrained on the proportion of foreign assets they can hold in their portfolio.
3. *Market efficiency*: Some foreign markets may exhibit liquidity problems. Another issue in market inefficiency is price manipulation. If foreign markets were too inefficient a manager would probably not run the risk of investing in these markets to benefit the domestic speculator.

Table 12.2► Transaction costs for the purchase of commercial property in selected countries

Country	Transaction costs
UK	3%
Ireland	6%
France	18.4%
Germany	2%
Portugal	10%
Australia	5.05%
USA	Varies across States and counties
Japan	9%
Singapore	3%

Sources: Dumortier (1997) and Morling (1997).

4. *Risk perception*: Foreign markets are perceived as risky by investors who are not familiar with them. Currency risk is also often perceived as an obstacle to international diversification.
5. *Costs*: The costs of international investments tend to be higher than those of domestic investments.

International investment also involves political risk such as political instability, the threat of nationalisation or restrictions on the repatriation of profits. The difficulties, which apply to international investments in general, are also valid for international property investments. These risks, however, are becoming less important. Local market knowledge constitutes one of the hurdles to international property investment. It is impossible for investors to have the same detailed knowledge they have of their domestic markets, for all or even most overseas markets. In the London office market in the 1980s, for instance, there was a common perception that many overseas investors were paying too much for property and that this was a result of lack of detailed market knowledge. The same problem arises for all investors considering investment outside their domestic markets. Lack of knowledge results in specific risk which is unrewarded. One way to overcome the problems is through joint ventures with local investors.

Different countries have different institutional contexts which require consideration. These result in differences in the following:

1. *Legal arrangements for land and property purchase*: These vary and specialist knowledge is required.
2. *Transaction costs*: Table 12.2 shows the substantial variation in transaction costs across countries.
3. *Leases*: These differences pertain to lease length, rent review, rent indexation, responsibilities and tenant right to cancel and renew. Table 12.3 summarises the main features of office property leases in the UK, France, Germany, Italy, Portugal, Australia and the USA.
4. *Planning systems*: The easier it is to develop, the higher the risk of oversupply and so of sharp falls in return.

Table 12.3► Characteristics of office property leases in selected countries

Country	Lease length (years)	Rent review (years)	Basis for rent indexation	Statutory renewal right	Ordinary repairs	Structural repairs
UK	25	5	Open market value (upward only)	Yes	Tenant	Tenant
France	9	1 or 3	Usually cost of building construction	Yes	Tenant	Landlord
Germany	5–10	Variable	Open market value or cost-of-living index	No	Tenant	Landlord
Italy	6	Annual	75% of retail price index	Yes	Tenant	Landlord
Portugal	5	None	—	Yes	Landlord with recovery	Landlord
Australia	3–5	Annual or mid-term	Open market value/Price index	No	Tenant	Landlord
USA	5–10	Variable	Variable	No	Landlord with recovery	Landlord

Sources: Arnold and Grossman (1995), Dubben and Sayce (1991), Gelbtuch *et al.* (1997), Morling (1997) and Worzala and Bernasek (1996).

5. *Tax regimes*: The VAT for new buildings, for instance, is 17.5% in the UK, 20.6% in France, 15% in Germany, 16% in Spain, 21% in Belgium, 3% in Singapore and 0.2% in Russia (Morling, 1997).
6. *Culture and negotiating methods*: These may differ substantially and what is acceptable or expected in the USA is not necessarily accepted or expected in Japan.

Another problem is that Modern Portfolio Theory (MPT, see Chapter 7) may suggest a large portfolio weight in a relatively small country. For large funds, there may not be enough property in the market to meet their requirements. In any case, in many overseas markets, even in capital cities, lot sizes are often much smaller than in major centres, such as London, and a large investor might be forced to buy a number of small properties making the process of international diversification difficult.

One of the questions asked by Worzala (1994) in her survey relates to attitudes of investors towards common problems often used to justify leaving overseas property out of an investment portfolio. Problems that respondents were asked to rank included:

▸ ability to identify acquisitions in a foreign market
▸ additional uncertainty due to currency fluctuations
▸ management and operation once investment is made
▸ taxation differences
▸ increased transaction costs
▸ lack of local expertise
▸ potential for misunderstandings due to cultural or language difficulties.

Most respondents considered the lack of local expertise, taxation differences and the identifying of overseas acquisitions as being the major problems. On the other hand, the investors surveyed were less sensitive to increased transaction costs and uncertainty caused by currency fluctuations. These results suggest that, at least at the time of the survey, investors were more concerned with implementing an international investment strategy rather than the possible consequences of making foreign investments.

One of the issues international investors are faced with is that of hedging currency risk. Unfortunately that part of Worzala's (1994) survey was not completed by most of the respondents, which suggests a lack of knowledge or interest in this part of the international investment decision. When hedging was listed, it was typically used to hedge only income cash flow, not entire investments.

12.6 Summary and conclusions

In Chapter 10, it was shown that property can play an important role in diversifying mixed-asset portfolios: the return on a portfolio can be increased (for a given risk level) or the risk can be diminished (for a given return). This chapter investigated whether the investment spectrum should be increased even further by considering international property investments. Intuitively such a strategy is appealing as country-specific risk should be diversified away. Other reasons for going international include differing lease structures or tax rates across countries.

The chapter first reviewed equilibrium relationships that should apply to product prices, interest rates and exchange rates. Five theoretical economic relationships were reviewed: purchasing power parity, the Fisher effect, the international Fisher effect, interest rate parity and forward rates as unbiased estimators of future spot rates. A basic understanding of these relationships is needed to manage international portfolios.

The main reasons for going international were reviewed. These include risk reduction, matching assets with liabilities and the limited size of the domestic property market. Risk reduction is the most established and frequently invoked argument in favour of international diversification. In some cases, international diversification even makes it possible to increase return and decrease risk. International strategies, however, imply exposure to currency risk, and the risk reduction resulting from the less than perfect correlation between assets across countries should be balanced with the additional risk (if any) due to currency movements. If an investor does not want to be exposed to currency risk, the investment may be hedged for major currencies.

The empirical evidence for international diversification was first examined for stocks and bonds. Then, empirical studies which have examined the benefits of international property diversification were reviewed. These studies generally show that international – direct or indirect – property investments make it possible to shift the efficient frontier upwards. In general, studies show that, when hedging of foreign currency is taken into account, international diversification is not attractive in a mean-variance portfolio framework. These studies, however, are not very reliable as the case of foreign property in one overseas country only is considered. If a full international diversification strategy is carried out, the risk reduction benefits should be substantially greater. Moreover, when several currencies are considered, hedging strategies may not be needed as currency risk should be diversified.

International property investment involves political risk such as political instability, the threat of nationalisation or restrictions on the repatriation of profits. These risks, however, are becoming less important. Barriers to foreign investment exist, but are usually exaggerated. These barriers include familiarity with foreign markets, regulations, market efficiency, risk perception and costs. An international property strategy also requires consideration of legal arrangements, transaction costs, lease structures, planning systems, tax regimes and culture and negotiating methods which vary across countries. With a large international portfolio, however, risks should be diversified and the benefits of an international strategy should clearly outweigh the disadvantages of such a strategy.

Notes

1. This section is based on Clark *et al.* (1993, Chapter 3), Shapiro (1996, Chapter 7) and Solnik (1993, Chapter 1).
2. As discussed in Chapter 1, shares and property are real assets, while bonds are financial assets.
3. The exchange rate can be a spot rate or a forward rate. The spot exchange rate is the rate of exchange of two currencies. The forward exchange rate is the rate of exchange of two currencies set on one date for delivery at a future specified date (for details see Clark *et al.*, 1993, Chapter 6).
4. There is always a bid price and an ask price for currencies. The bid price is the price at which the bank is willing to buy the currency and the ask price is the price at which the bank is willing to sell the currency. Differences always exist between the bid and ask prices because of transactions costs. For widely traded currencies the spread will be lower than for less heavily traded currencies.
5. Short-run deviations could occur because price indices are constructed in different ways across countries (the goods and services included in the market basket and the weighting formula are different) and because prices are sticky in the short run.
6. The notation used in this chapter is consistent with that used in the finance literature, but is slightly different from the one used in Chapters 2 and 5.
7. As defined by Shapiro (1996: 252): 'hedging a particular currency exposure means establishing an offsetting currency position such that whatever is lost or gained on the original currency exposure is exactly offset by a corresponding foreign exchange gain or loss on the currency hedge.'
8. See Hamelink *et al.* (1999) for more details on the use of cluster analysis to form homogeneous property market groups.
9. As stated by Worzala (1995: 26): 'the various hedging techniques available to mitigate exchange rate fluctuations for an international exposure are meant to be used for a specified period of time to hedge known cash inflows or outflows against adverse movements in the exchange rate.' For long-term investments, such as property investments, the forward contracts used by Worzala in her study will have to be rolled over at the end of each contract period.
10. Property shares have been shown to react partly to the same factors which affect common stocks and to be hybrid assets of common stocks and property. As property

share international diversification is more effective than for common stocks, it is reasonable to think that direct property diversification should be even more effective as the common stock factors would be eliminated. To date no such study has been undertaken, primarily because of data limitations.

Further reading

Ball, M., Lizieri, C. and MacGregor, B. D. (1998) *The economics of commercial property markets*, Routledge, London: Chapter 13.

Baum, A. E. (1995) Can foreign real estate investment be successful?, *Real Estate Finance*, 12(1), 81–9.

Brueggeman, W. B. and Thibodeau, T. G. (1997) *Investment performance and benefits from global diversification and real estate equity securities: a preliminary analysis*, paper presented at the AREUEA 6th international real estate conference, University of California at Berkeley.

Clark, E., Levasseur, M. and Rousseau, P. (1993) *International finance*, Chapman & Hall, London: Chapters 3, 6, 8, 10 and 11.

Eichholtz, P. M. A. (1997) Real estate securities and common stocks: a first international look, *Real Estate Finance*, 14(1), 70–4.

Eichholtz, P. M. A. and Koedijk, K. G. (1996) International real estate securities indexes, *Real Estate Finance*, 12(4), 42–50.

Eichholtz, P. M. A. and Koedijk, K. G. (1996) The global real estate securities market, *Real Estate Finance*, 13(1), 76–82.

Journal of Property Valuation and Investment (1997) Special issue on European real estate investment, Lizieri, C. (Guest editor), 15(4).

Lizieri, C. and Finlay, L. (1995) International property portfolio strategies: problems and opportunities, *Journal of Property Valuation and Investment*, 13(1), 6–21.

Real Estate Economics (1995) Special issue on European housing and mortgage markets, Dale-Johnson, D. and Gabriel, S. A. (Guest editors), 23(4).

Shapiro, A. C. (1996) *Multinational financial management*, Prentice Hall International, Inc., Upper Saddle River (NJ) (fifth edition): Chapters 6, 7 and 16.

Solnik, B. (1993) *International investments*, Addison Wesley, Reading (MA) (second edition): Chapters 1, 2 and 10.

The Journal of Real Estate Finance and Economics (1993) Special issue on the Asian real estate market, 6(1).

The Journal of Real Estate Finance and Economics (1997) Special issue on European real estate, Hoesli, M. and MacGregor, B. D. (Guest editors), 15(1).

The Journal of Real Estate Research (1996) Special issue on international real estate investment, 11(2).

The Journal of Real Estate Research (1997) Special issue on international real estate investment, 13(3).

Wurtzebach, C. H. and Baum, A. E. (1994) International real estate, in *Managing real estate portfolios*, Hudson-Wilson, S. and Wurtzebach, C. H. (Eds), Irwin, Burr Ridge (IL), 284–308: Chapter 8.

Ziobrowski, B. and Ziobrowski, A. (1995) Using forward contracts to hedge investment in US real estate, *Journal of Property Valuation and Investment*, 13(1), 22–43.

References

Abraham, J. M. and Schauman, W. S. (1991) New evidence on home prices from Freddie Mac repeat sales, *Journal of the American Real Estate and Urban Economics Association*, 19(3), 333–52.

Adair, A., Hutchison, N., MacGregor, B. D., McGreal, S. and Nanthakumaran, N. (1996) Variations in the capital valuations of UK commercial property, *Journal of Property Valuation and Investment*, 14(5), 34–47.

Adams, A. (1989) *Investment*, Graham & Trotman, London.

Ahern, T., Liang, Y. and Myer, F. C. N. (1998) Leverage in a pension fund real estate program, *Real Estate Finance*, 15(2), 55–62.

Alexander, G. J. and Francis, J. C. (1986) *Portfolio analysis*, Prentice Hall, New Jersey (third edition).

Arnold, H. R. and Grossman, C. (1995) International real estate investment: a realistic look at the issues, in *The handbook of real estate portfolio management*, Pagliari, J. L. Jr (Ed.), Irwin, Burr Ridge (IL), 530–67.

Bailey, M. J., Muth, R. F. and Nourse, H. O. (1963) A regression method for real estate price index construction, *Journal of the American Statistical Association*, 58, 933–42.

Ball, M., Lizieri, C. and MacGregor, B. D. (1998) *The economics of commercial property markets*, Routledge, London.

Barkham, R. (1995) The performance of the UK property company sector: a guide to investing in British property via the public stock market, *Real Estate Finance*, 12(1), 90–98.

Barkham, R. and Geltner, D. (1994) Unsmoothing British valuation-based returns without assuming an efficient market, *Journal of Property Research*, 11(2), 81–95.

Barkham, R. J. and Geltner, D. M. (1995) Price discovery in American and British property markets, *Real Estate Economics*, 23(1), 21–44.

Barkham, R. J. and Ward, C. W. R. (1999) Investor sentiment and noise traders: discount to net asset value in listed property companies in the UK, *Journal of Real Estate Research*, 18(2), 291–312.

Barnett, G. and Rosenberg, M. (1983) International diversification in bonds, *Prudential international fixed income investment strategy*, second quarter.

Barras, R. (1983) A simple theoretical model of the office development cycle, *Environment and Planning A*, 15, 1361–94.

Barras, R. and Clark, P. (1996) Obsolescence and performance in the Central London market, *Journal of Property Valuation and Investment*, 14(4), 63–78.

Barras, R. and Ferguson, D. (1987a) Dynamic modelling of the building cycle: 1. Theoretical framework, *Environment and Planning A*, 19, 353–67.

Barras, R. and Ferguson, D. (1987b) Dynamic modelling of the building cycle: 2. Empirical results, *Environment and Planning A*, 19, 493–520.

Baum, A. E. (1991) *Property investment depreciation and obsolescence*, Routledge, London.

Baum, A. E. (1998) Depreciation and property investment appraisal, in *Property investment theory*, MacLeary, A. and Nanthakumaran, N. (Eds), E & FN Spon, London, 48–69.

Baum, A. E. and Crosby, N. (1995) *Property investment appraisal*, Routledge, London (second edition).

Baum, A. E., Crosby, N. and MacGregor, B. D. (1996) Price formation, mispricing and investment analysis in the property market, *Journal of Property Valuation and Investment*, 14(1), 36–49.

Baum, A. E. and Lee, S. (1990) *Foundations of property investment*, Department of Land Management and Development, University of Reading.

Baum, A. and Lizieri, C. (1998) *Who owns the City?* Development Securities/University of Reading, London.

Baum, A. E. and MacGregor, B. D. (1992) The initial yield revealed: explicit valuations and the future of property investment, *Journal of Property Valuation and Investment*, 10(4), 709–27.

Baum, A. E. and Schofield, A. (1991) Property as a global asset, in *Investment, procurement and performance in construction*, Venmore-Rowland, P., Brandon, P. and Mole, T. (Eds), E & FN Spon, London, 103–55.

Bender, A. R., Gacem, B. and Hoesli, M. (1994) Construction d'indices immobiliers selon l'approche hédoniste, *Financial markets and portfolio management*, 8(4), 522–34.

Benjamin, J. D., Judd, D. and Winkler, D. T. (1995) An analysis of shopping center investment, *The Journal of Real Estate Finance and Economics*, 10(2), 161–8.

Blundell, G. F. and Ward, C. W. R. (1987) Property portfolio allocation: a multi-factor model, *Land Development Studies*, 4, 145–56.

Bowie, N. (1983) The depreciation of buildings, *Journal of Valuation*, 2(1), 5–13.

Boykin, J. H. and Ring, A. A. (1993) *The valuation of real estate*, Regents/Prentice Hall, Englewood Cliffs (fourth edition).

Brown, G. R. (1985) The information content of property valuations, *Journal of Valuation*, 3, 350–62.

Brown, G. R. (1991) *Property investment and the capital markets*, E & FN Spon, London.

Brown, G. R. (1992) Valuation accuracy: developing the economic issues, *Journal of Property Research*, 9, 199–207.

Brown, G. R. and Schuck, E. J. (1996) Optimal portfolio allocations to real estate, *Journal of Real Estate Portfolio Management*, 2(1), 63–73.

Brueggeman, W. B. and Fisher, J. D. (1997) *Real estate finance and investments*, Irwin, Chicago (tenth edition).

Büchel, S. and Hoesli, M. (1995) A hedonic analysis of rent and rental revenue in the subsidised and unsubsidised housing sectors in Geneva, *Urban Studies*, 32(7), 1199–213.

Case, B., Goetzmann, W. N. and Wachter, S. M. (1997) *The global commercial property market cycles: a comparison across property types*, paper presented at the AREUEA 6th international real estate conference, University of California at Berkeley.

Case, B. and Quigley, J. M. (1991) The dynamics of real estate prices, *The Review of Economics and Statistics*, 73(1), 50–58.

Case, K. and Shiller, R. (1989) The efficiency of the market for single family homes, *American Economic Review*, 79(1), 125–37.

Case, K. E., Shiller, R. J. and Weiss, A. N. (1993) Index-based futures and options markets in real estate, *The Journal of Portfolio Management*, Winter, 83–92.

Chan, K. C., Hendershott, P. H. and Sanders, A. B. (1990) Risk and return on real estate: evidence from equity REITs, *Journal of the American Real Estate and Urban Economics Association*, 18(4), 431–52.

Chaplin, R. (1997) Unsmoothing valuation-based indices using multiple regimes, *Journal of Property Research*, 14(3), 189–210.

Chau, K. W., MacGregor, B. D. and Schwann, G. M. (1998) *Price discovery in the Hong Kong real estate market*, paper presented at the European Real Estate Society/American Real Estate and Urban Economics Association conference, Maastricht, The Netherlands, 10–13 June.

Chun, G. H. and Shilling, J. D. (1998) Real estate asset allocations and international real estate markets, *Journal of the Asian Real Estate Society*, 1(1), 17–44.

Clapp, J. and Giaccotto, C. (1992) Estimating price indices for residential property: a comparison of repeat sales and assessed value methods, *Journal of the American Statistical Association*, 87, 300–306.

Clapp, J. M., Giaccotto, C. and Tirtiroglu, D. (1991) Housing price indices based on all transactions compared to repeat subsamples, *Journal of the American Real Estate and Urban Economics Association*, 19(3), 270–85.

Clark, E., Levasseur, M. and Rousseau, P. (1993) *International finance*, Chapman & Hall, London.

Cole, R. A. (1988) *A new look at commercial real estate returns*, PhD dissertation, The University of North Carolina at Chapel Hill.

Coleman, M., Hudson-Wilson, S. and Webb, J. R. (1994) Real estate in the multi-asset portfolio, in *Managing real estate portfolios*, Hudson-Wilson, S. and Wurtzebach, C. H. (Eds), Irwin, Burr Ridge (IL), 98–123.

Corgel, J. B. and Gay, G. D. (1987) Local economic base, geographic diversification and risk management of mortgage portfolios, *Journal of the American Real Estate and Urban Economics Association*, 15(3), 256–67.

Corgel, J. B., Smith, H. C. and Ling, D. C. (1998) *Real estate perspectives: an introduction to real estate*, Irwin/McGraw-Hill, Boston (third edition).

Court, A. (1939) Hedonic price indexes with automotive examples, in *The dynamics of automobile demand*, General Motors Corporation, New York, 99–117.

Crone, T. M. and Voith, R. P. (1992) Estimating house price appreciation: a comparison of methods, *Journal of Housing Economics*, 2, 324–38.

De Wit, D. P. M. (1996) Real estate portfolio management practices of pension funds and insurance companies in the Netherlands: a survey, *The Journal of Real Estate Research*, 11(2), 131–48.

Dohrmann, G. (1995) The evolution of institutional investment in real estate, in *The handbook of real estate portfolio management*, Pagliari, J. L. Jr (Ed.), Irwin, Burr Ridge (IL), 3–116.

Dombrow, J., Knight, J. R. and Sirmans, C. F. (1997) Aggregation bias in repeat-sales indices, *The Journal of Real Estate Finance and Economics*, 14(1/2), 75–88.

Downs, D. H. and Hartzell, D. J. (1995) Real estate investment trusts, in *The handbook of real estate portfolio management*, Pagliari, J. L. Jr (Ed.), Irwin, Burr Ridge (IL), 597–634.

Drivers Jonas/IPD (1988) *The variation in valuations*, Drivers Jonas/IPD, London.

Drivers Jonas/IPD (1990) *The variation in valuations – 1990 update*, Drivers Jonas/IPD, London.

Dubben, N. and Sayce, S. (1991) *Property portfolio management: an introduction*, Routledge, London.

Dumortier, J.-P. (1997) Le nouveau contexte économique et financier: quelle part d'allocation d'actifs dans l'immobilier?, *Réflexions immobilières*, October, 12–18.

Dunse, N. and Jones, C. (1998) A hedonic price model of office rents, *Journal of Property Valuation and Investment*, 16(3), 297–312.

Eichholtz, P. M. A. (1996) Does international diversification work better for real estate than for stocks and bonds? *Financial Analysts Journal*, January–February, 56–62.

Eichholtz, P. M. A. (1997a) A long run house price index: The Herengracht index, 1628–1973, *Real Estate Economics*, 25(2), 175–92.

Eichholtz, P. M. A. (1997b) Real estate securities and common stocks: a first international look, *Real Estate Finance*, 14(1), 70–74.

Eichholtz, P. M. A., de Graaf, N., Kastrop, W. and Op 't Veld, H. (1998) Introducing the GPR 250 property share index, *Real Estate Finance*, 15(1), 51–61.

Eichholtz, P. M. A. and Hartzell, D. J. (1996) Property shares, appraisals and the stock market: an international perspective, *The Journal of Real Estate Finance and Economics*, 12(2), 163–78.

Eichholtz, P. M. A., Hoesli, M., MacGregor, B. D. and Nanthakumaran, N. (1995) Real estate portfolio diversification by property type and region, *Journal of Property Finance*, 6(3), 39–59.

Eichholtz, P., Huisman, R., Koedijk, K. and Schuin, L. (1998) Continental factors in international real estate returns, *Real Estate Economics*, 26(3), 493–509.

Eichholtz, P. M. A. and Koedijk, K. G. (1996a) International real estate securities indexes, *Real Estate Finance*, 12(4), 42–50.

Eichholtz, P. M. A. and Koedijk, K. G. (1996b) The global real estate securities market, *Real Estate Finance*, 13(1), 76–82.

Eichholtz, P., Koedijk, K. and Schweitzer, M. (1997) *Testing international real estate investment strategies*, paper presented at the AREUEA 6th international real estate conference, University of California at Berkeley.

Eichholtz, P. M. A., Mahieu, R. J. and Schotman, P. S. (1993) *Real estate diversification: by country or by continent?*, Working paper, Limburg University, Maastricht.

Ennis, R. M. and Burik, P. (1991) Pension fund real estate investment under a simple equilibrium pricing model, *Financial Analysts Journal*, 47(3), 20–30.

Fama, E. F. (1970) Efficient capital markets: a review of theory and empirical work, *Journal of Finance*, 25, 383–420.

Field, B. G. and MacGregor, B. D. (1987) *Forecasting techniques in urban and regional planning*, Hutchinson, London.

Firstenberg, P. B., Ross, S. A. and Zisler, R. C. (1988) Real estate: the whole story, *The Journal of Portfolio Management*, 14(3), 22–34.

Fisher, J. D., Geltner, D. M. and Webb, R. B. (1994) Value indices of commercial real estate: a comparison of index construction methods, *The Journal of Real Estate Finance and Economics*, 9(2), 137–64.

Flanagan, R., Norman, G., Meadows, J. and Robinson, G. (1989) *Life cycle costing: theory and practice*, BSP Professional Books, London.

Fogler, H. R. (1984) 20% in real estate: can theory justify it?, *The Journal of Portfolio Management*, 10(2), 6–13.

Follain, J. R. and Calhoun, C. A. (1997) Constructing indices of the price of multi-family properties using the 1991 Residential Finance Survey, *The Journal of Real Estate Finance and Economics*, 14(1/2), 235–55.

Follain, J., Hendershott, P. H. and Ling, D. (1987) Understanding the real estate provisions of the Tax Act: their motivation and impact, *National Tax Journal*, 363–72.

Follain, J., Hendershott, P. H. and Ling, D. (1992) Real estate markets since 1980: what role have tax changes played?, *National Tax Journal*, 253–66.

Gardiner, C. and Henneberry, J. (1988) The development of a simple regional office rent prediction model, *Journal of Valuation*, 7, 36–52.

Gardiner, C. and Henneberry, J. (1991) Predicting regional office rents using habit-persistence theories, *Journal of Property Valuation and Investment*, 9, 215–26.

Gatzlaff, D. and Geltner, D. (1998) A transaction-based index of commercial property and its comparison to the NCREIF index, *Real Estate Finance*, 15(1), 7–22.

Gatzlaff, D. H. and Tirtiroglu, D. (1995) Real estate market efficiency: issues and evidence, *Journal of Real Estate Literature*, 3(2), 157–92.

Gelbtuch, H. C., Mackmin, D. and Milgrim, M. R. (Eds) (1997) *Real estate valuation in global markets*, Appraisal Institute, Chicago (IL).

Geltner, D. (1993a) Temporal aggregation in real estate return indices, *Journal of the American Real Estate and Urban Economics Association*, 21(2), 141–66.

Geltner, D. (1993b) Estimating market values from appraised values without assuming an efficient market, *The Journal of Real Estate Research*, 8(3), 325–45.

Ghosh, C., Miles, M. and Sirmans, C. F. (1996) Are REITs stocks?, *Real Estate Finance*, 13(3), 46–53.

Giliberto, M., Hamelink, F., Hoesli, M. and MacGregor, B. D. (1999) Optimal diversification within mixed-asset portfolios using a conditional heteroscedasticity approach: evidence from the US and the UK, *Journal of Real Estate Portfolio Management*, 5(1), 31–45.

Giliberto, M. and Mengden, A. (1996) REITs and real estate: two markets re-examined, *Real Estate Finance*, 13(1), 56–60.

Giliberto, S. M. and Sidoroff, F. N. (1995) Real estate stock indexes, *Real Estate Finance*, 12(1), 56–62.

Goetzmann, W. N. and Ibbotson, R. G. (1990) The performance of real estate as an asset class, *Journal of Applied Corporate Finance*, 3(1), 65–76.

Gordon, J. N., Canter, T. A. and Webb, J. R. (1998) The effect of international real estate securities on portfolio diversification, *Journal of Real Estate Portfolio Management*, 4(2), 83–92.

Grenadier, S. R. (1995) Valuing lease contracts: a real options approach, *Journal of Financial Economics*, 38, 297–331.

Griliches, Z. (1961) Hedonic price indexes for automobiles: an econometric analysis of quality change, in *The price statistics of the federal government*, General series, No. 73, National Bureau of Economic Research, New York, 137–96.

Gujarati, D. (1992) *Essentials of econometrics*, McGraw-Hill International Editions, New York.

Hamelink, F., Hoesli, M. and MacGregor, B. (1997) Inflation hedging versus inflation protection in the US and the UK, *Real Estate Finance*, 14(2), 63–73.

Hamelink, F., Hoesli, M., Lizieri, C. and MacGregor, B. D. (1999) Homogeneous commercial property market groups and portfolio construction in the UK, *Environment and Planning A*, (forthcoming).

Hamelink, F., MacGregor, B. D., Nanthakumaran, N. and Orr, A. M. (1998) A comparison of UK equity and property duration, *Aberdeen Papers in Land Economy*, 98–103.

Han, J. (1996) To securitize or not to securitize? The future of commercial real estate debt markets, *Real Estate Finance*, 13(2), 71–80.

Han, J. and Liang, Y. (1995) The historical performance of real estate investment trusts, *The Journal of Real Estate Research*, 10(3), 235–62.

Hartzell, D. J., Eichholtz, P. and Selender, A. (1993) Economic diversification in European real estate portfolios, *Journal of Property Research*, 10(1), 5–25.

Hartzell, D. J., Hekman, J. and Miles, M. (1986) Diversification categories in investment real estate, *Journal of the American Real Estate and Urban Economics Association*, 14(2), 230–54.

Hartzell, D. J., Watkins, D. E. and Laposa, S. P. (1997) *Performance characteristics of global real estate securities*, paper presented at the AREUEA annual meeting, New Orleans.

Hendershott, P. H. (1981) The decline in aggregate share values: taxation, valuation errors, risk and profitability, *American Economic Review*, 909–22.

Hendershott, P. H. (1995) Real effective rent determination: evidence from the Sydney office market, *Journal of Property Research*, 12(2), 127–35.

Hendershott, P. H. (1997) Uses of equilibrium models in real estate research, *Journal of Property Research*, 14(1), 1–13.

Hendershott, P. H., Follain, J. and Ling, D. (1987) Effects on real estate, in *Tax Reform and the U.S. Economy*, Pechman, J. (ed.), The Brookings Institution, Washington DC, 71–94.

Hendershott, P. H. and Ling, D. (1984a) Prospective changes in tax law and the value of depreciable real estate, *Journal of the American Real Estate and Urban Economics Association*, 12(3), 297–317.

Hendershott, P. H. and Ling, D. (1984b) Trading and the tax shelter value of depreciable real estate, *National Tax Journal*, 213–23.

Hendershott, P. H., Lizieri, C. and Matysiak, G. A. (1999) The workings of the London office market, *Real Estate Economics*, 27(2), 365–87.

Hetherington, J. (1988) Forecasting of rents, in *Property investment theory*, MacLeary, A. and Nanthakumaran, N. (Eds), E & FN Spon, London, 97–107.

Hill, R. C., Knight, J. R. and Sirmans, C. F. (1997) Estimating capital asset price indexes, *The Review of Economics and Statistics*, 79(2), 226–33.

Hoag, J. W. (1980) Towards indices of real estate value and return, *The Journal of Finance*, 35(2), 569–80.

Hoesli, M. (1993) *Investissement immobilier et diversification de portefeuille*, Economica, Paris.

Hoesli, M. (1998) L'accession à la propriété à Genève: problèmes et mesures, *Réflexions immobilières*, July, 48–54.

Hoesli, M. E. and Anderson, M. S. (1991) Swiss real estate: return, risk and diversification opportunities, *Journal of Property Research*, 8(2), 133–45.

Hoesli, M., Giaccotto, C. and Favarger, P. (1997) Three new real estate price indices for Geneva, Switzerland, *The Journal of Real Estate Finance and Economics*, 15(1), 93–109.

Hoesli, M. and Hamelink, F. (1996) Diversification of Swiss portfolios with real estate: results based on a hedonic index, *Journal of Property Valuation and Investment*, 14(5), 59–75.

Hoesli, M. and Hamelink, F. (1997) An examination of the role of Geneva and Zurich housing in Swiss institutional portfolios, *Journal of Property Valuation and Investment*, 15(4), 354–71.

Hoesli, M., Lizieri, C. M. and MacGregor, B. D. (1997) The spatial dimensions of the investment performance of UK commercial property, *Urban Studies*, 34(9), 1475–94.

Hoesli, M. and MacGregor, B. (1995) The classification of local property markets in the UK using cluster analysis, in *The Cutting Edge: Proceedings of the RICS property research conference 1995*, Vol. 1, RICS, London, 185–204.

Hoesli, M., MacGregor, B. D., Matysiak, G. A. and Nanthakumaran, N. (1997) The short-term inflation-hedging characteristics of UK real estate, *The Journal of Real Estate Finance and Economics*, 15(1), 27–58.

Hoesli, M. and Thion, B. (1994) *Immobilier et gestion de partrimoine*, Economica, Paris.

Hough, D. E. and Kratz, C. G. (1983) Can good architecture meet the market test?, *Journal of Urban Economics*, 14, 40–54.

Hutchison, N. E. (1994) Housing as an investment? A comparison of returns from housing with other types of investment, *Journal of Property Finance*, 5(2), 47–61.

Ibbotson, R. G. and Siegel, L. B. (1984) Real estate returns: a comparison with other investments, *Journal of the American Real Estate and Urban Economics Association*, 12(3), 219–42.

ICPA – The Commercial Network (1996) *International property survey.*

IPD (1997) *IPD Property investors' digest 1997*, Investment Property Databank, London.

Jackson, C. (1998) Developing a classification of local property markets on the basis of retail rental growth rates, *Aberdeen Papers in Land Economy*, 98–101.

Kallberg, J. G., Liu, C. H. and Greig, D. W. (1996) The role of real estate in the portfolio allocation process, *Real Estate Economics*, 24(3), 359–77.

Key, T., MacGregor, B. D., Nanthakumaran, N. and Zarkesh, F. (1994) *Economic cycles and property cycles*, Main report for Understanding the property cycle, Royal Institution of Chartered Surveyors, London.

Khalid, A. G. (1992) *Hedonic price estimation of the financial impact of obsolescence on commercial office buildings*, Unpublished PhD thesis, University of Reading.

Khoo, T., Hartzell, D. and Hoesli, M. (1993) An investigation of the change in real estate investment trusts betas, *Journal of the American Real Estate and Urban Economics Association*, 21(2), 107–30.

Kiel, K. A. and Zabel, J. E. (1997) Evaluating the usefulness of the American Housing Survey for creating house price indices, *The Journal of Real Estate Finance and Economics*, 14(1/2), 189–202.

Knight, J. R., Dombrow, J. and Sirmans, C. F. (1995) A varying parameters approach to constructing house price indexes, *Real Estate Economics*, 23(2), 187–205.

Lambert Smith Hampton and HRES (1997) *Trophy or tombstone?: a decade of depreciation in the Central London office market*, Lambert Smith Hampton and HRES, London.

Lancaster, K. J. (1966) A new approach to consumer theory, *Journal of Political Economy*, 74, 132–57.

Lee, S., Byrne, P. and French, N. (1996) Assessing the future of property in the multi-asset portfolio: the case of UK pension funds, *Journal of Property Research*, 13(3), 197–209.

Leggett, D. N. (1995) An empirical analysis of efficient real property liquidity premiums, in *Alternative ideas in real estate investment*, Schwarz, A. L. and Kapplin, S. D. (Eds), sponsored by the American Real Estate Society, Kluwer Academic Publishers, Boston, 113–27.

Liang, Y., Chatrath, A. and McIntosh, W. (1996) Apartment REITs and apartment real estate, *The Journal of Real Estate Research*, 11(3), 277–89.

Lin, C. C. S. (1993) The relationship between rents and prices of owner-occupied housing in Taiwan, *The Journal of Real Estate Economics and Finance*, 6(1), 25–54.

Ling, D. C. and Naranjo, A. (1998) The fundamental determinants of commercial real estate returns, *Real Estate Finance*, 14(4), 13–24.

Lintner, J. (1965) The valuation of risk assets and the selection of risky investments in stock portfolios and capital budgets, *The Review of Economics and Statistics*, 47, 13–37.

Liu, C. H., Grissom, T. V. and Hartzell, D. J. (1995) Superior real estate performance: enigma or illusion? A critical review of the literature, in *Alternative ideas in real estate investment*, Schwarz, A. L. and Kapplin, S. D. (Eds), sponsored by the American Real Estate Society, Kluwer Academic Publishers, Boston, 59–82.

Liu, C. H., Hartzell, D. J., Grissom, T. V. and Grieg, W. (1990) The composition of the market portfolio and real estate investment performance, *Journal of the American Real Estate and Urban Economics Association*, 18(1), 49–75.

Liu, C. H. and Mei, J. (1998) The predictability of international real estate markets, exchange rate risks and diversification consequences, *Real Estate Economics*, 26(1), 3–39.

Lizieri, C. and Venmore-Rowlands, P. (1991) Valuation accuracy: a contribution to the debate, *Journal of Property Research*, 8(2), 115–22.

Lizieri, C. and Venmore-Rowlands, P. (1993) Valuations prices and the market: a rejoinder, *Journal of Property Research*, 10(2), 77–84.

Lumby, S. (1988) *Investment appraisal and financing decisions*, VNR International, London (third edition).

MacGregor, B. D. and Nanthakumaran, N. (1992) The allocation to property in the multi-asset portfolio: the evidence and theory reconsidered, *Journal of Property Research*, 9(1), 5–32.

MacGregor, B. D. and Schwann, G. (1999) *Common features in UK commercial property market returns*, paper presented at the Pacific Rim Real Estate Society Annual Conference, Kuala Lumpur, Malaysia, January.

Malizia, E. E. and Simons, R. A. (1991) Comparing regional classifications for real estate portfolio diversification, *The Journal of Real Estate Research*, 6(1), 53–67.

Mark, J. H. and Goldberg, M. A. (1984) Alternative housing price indices: an evaluation, *Journal of the American Real Estate and Urban Economics Association*, 12(1), 30–49.

Markowitz, H. M. (1952) Portfolio selection, *The Journal of Finance*, 7(1), 77–91.

Markowtiz, H. M. (1959) *Portfolio selection: efficient diversification of investments*, John Wiley & Sons, New York.

Matysiak, G. A. and Wang, P. (1995) Commercial property market prices and valuations: analysing the correspondence, *Journal of Property Research*, 12(3), 181–202.

Meese, R. A. and Wallace, N. E. (1997) The construction of residential housing price indices: a comparison of repeat-sales, hedonic-regression, and hybrid approaches, *The Journal of Real Estate Finance and Economics*, 14(1/2), 51–73.

Mei, J. and Saunders, A. (1997) Have US financial institutions' real estate investments exhibited 'trend-chasing' behavior?, *The Review of Economics and Statistics*, 79(2), 248–58.

Miles, M., Cole, R. and Guilkey, D. (1990) A different look at commercial real estate returns, *Journal of the American Real Estate and Urban Economics Association*, 18(4), 403–30.

Miles, M., Roberts, J., Machi, D. and Hopkins, R. (1994) Sizing the investment markets: a look at the major components of public and private markets, *Real Estate Finance*, 11(1), 39–50.

Morling, I. (1997) La France et sa compétitivité mondiale: un survol mondial, *Réflexions immobilières*, April, 76–84.

Morrell, G. D. (1991) Property performance analysis and performance indices: a review, *Journal of Property Research*, 8(1), 29–57.

Mossin, J. (1966) Equilibrium in a capital asset market, *Econometrica*, 34(4), 768–83.

Mueller, G. R. (1993) Redefining economic diversification strategies for real estate portfolios, *The Journal of Real Estate Research*, 8(1), 55–68.

Mueller, G. R. and Laposa, S. P. (1996) REIT returns: a property-type perspective, *Real Estate Finance*, 13(1), 45–55.

Mueller, G. R. and Ziering, B. A. (1992) Real estate portfolio diversification using economic diversification, *The Journal of Real Estate Research*, 7(4), 375–86.

Mueller, G. R., Pauley, K. R. and Morrill, W. K. (1994) Should REITs be included in a mixed-asset portfolio?, *Real Estate Finance*, 11(1), 23–8.

Newell, G. and Brown, G. (1994) *Correcting for appraisal smoothing in real estate returns*, working paper, Property Research Centre, University of Western Sydney.

Newell, G. and Chau, K. W. (1996) Linkages between direct and indirect property performance in Hong Kong, *Journal of Property Finance*, 7(4), 9–29.

Newell, G. and MacFarlane, J. (1995) Improved risk estimation using appraisal-smoothed real estate returns, *Journal of Real Estate Portfolio Management*, 1(1), 51–7.

Newell, G. and MacFarlane, J. (1996) Risk estimation and appraisal-smoothing in UK property returns, *Journal of Property Research*, 13(1), 1–12.

Newell, G. and MacFarlane, J. (1998) The effect of seasonality of valuations on property risk, *Journal of Property Research*, 15(3), 167–82.

Newell, G. and Webb, J. R. (1995) Institutional real estate performance benchmarks: an international perspective, *Journal of Real Estate Literature*, 2(2), 215–23.

Newell, G. and Webb, J. (1996) Assessing risk for international real estate investments, *The Journal of Real Estate Research*, 11(2), 103–15.

Newell, G. and Webb, J. R. (1998) Real estate performance benchmarks in New Zealand and South Africa, *Journal of Real Estate Literature*, 6(2), 137–43.

Norman, E. J., Sirmans, G. S. and Benjamin, J. D. (1995) The historical environment of real estate returns, *Journal of Real Estate Portfolio Management*, 1(1), 1–24.

Pagliari, J. L. Jr, Lieblich, F., Schaner, M. and Webb, J. R. (1998) *Twenty years of the NCREIF property index*, working draft, SSR Realty Advisors.

Palmquist, R. B. (1982) Measuring environmental effects on property values without hedonic regressions, *Journal of Urban Economics*, 11, 333–47.

Quan, D. C. and Titman, S. (1997) Commerical real estate prices and stock market returns: an international analysis, *Financial Analysts Journal*, May–June, 21–34.

Quigley, J. M. (1995) A simple hybrid model for estimating real estate price indexes, *Journal of Housing Economics*, 4, 1–12.

Radcliffe, R. C. (1994) *Investment: concepts, analysis, strategy*, Harper Collins, New York (fourth edition).

Richard Ellis (1997) *World rental levels: offices*, July.

Riddiough, T. J. (1995) Replicating and hedging real estate risk, *Real Estate Finance*, 12(3), 88–95.

Rosen, S. (1974) Hedonic prices and implicit markets: product differentiation in pure competition, *Journal of Political Economy*, 82, 34–55.

Ross, S. A. and Zisler, R. C. (1991) Risk and return in real estate, *The Journal of Real Estate Finance and Economics*, 4(2), 175–90.

RICS (1995) *RICS appraisal and valuation manual*, Royal Institution of Chartered Surveyors, London. ('the Red Book')

Rutterford, J. (1993) *Introduction to stock exchange investment*, Macmillan, Basingstoke (second edition).

Salway, F. (1986) *Depreciation of commercial property*, CALUS, University of Reading.

Schärer, C. (1997) *Immobilienanlagen schweizerischer Pensionskassen*, Haupt, Bern.

Shapiro, A. C. (1996) *Multinational financial management*, Prentice Hall International, Inc., Upper Saddle River (NJ) (fifth edition).

Sharpe, W. F. (1964) Capital asset prices: a theory of market equilibrium under conditions of risk, *The Journal of Finance*, 19(3), 425–42.

Sharpe, W. F. (1990) Asset allocation, in *Managing investment portfolios*, Maginn, J. L. and Tuttle, D. L. (Eds), Warren Graham & Lamont (second edition).

Shiller, R. J. (1993) *Macro markets*, Clarendon Press, Oxford.

Shilling, J. D., Sirmans, C. F. and Corgel, J. B. (1987) Price adjustment process for rental office space, *Journal of Urban Economics*, 22, 90–100.

Shulman, D. and Hopkins, R. E. (1988) *Economic diversification in real estate portfolios*, Bond market research: real estate, Salomon Brothers Inc., New York, November.

Sirmans, C. F. (1980) Minimum tax, recapture and the choice of depreciation methods, *Journal of the American Real Estate and Urban Economics Association*, 8(3), 255–67.

Solnik, B. (1993) *International investments*, Addison-Wesley, Reading (MA) (second edition).

Solnik, B. H. and Noetzlin, B. (1982) Optimal international asset allocation, *The Journal of Portfolio Management*, Fall, 11–21.

StatSoft (1996) *Statistica for Windows: general conventions and statistics*, Volume 1, Statsoft Inc. (second edition).

Thibodeau, T. G. (1992) *Residential real estate prices: 1974–1983*, The Blackstone Company, Mount Pleasant.

Thion, B. and Riva, F. (1997) Performances des sociétés immobilières à la Bourse de Paris, *Banque & Marchés*, September–October, 31–46.

Tsolacos, S. (1998) Econometric modelling and forecasting of new retail development, *Journal of Property Research*, 15(4), 265–84.

Ward, C. W. R. (1988) Asset pricing models and property as a long-term investment, in *Property investment theory*, MacLeary, A. and Nanthakumaran, N. (Eds), E & FN Spon, London, 134–45.

Ward, C. W. R. (1999) Personal communication.

Ward, C. W. R., Hendershott, P. H. and French, N. S. (1998) Pricing upward-only rent review clauses: an international perspective, *Journal of Property Valuation and Investment*, 1998, 16, 447–54.

Webb, J. R. and Rubens, J. H. (1987) How much in real estate? A surprising answer, *The Journal of Portfolio Management*, 13(3), 10–14.

Webb, R. B. (1994) On the reliability of commercial appraisals, *Real Estate Finance*, 11(1), 62–5.

Webb, R. B., Miles, M. and Guilkey, D. (1992) Transactions-driven commercial real estate returns: the panacea to asset allocation models?, *Journal of the American Real Estate and Urban Economics Association*, 20(2), 325–57.

Wheaton, W. C. and Torto, R. G. (1990) An investment model of the demand and supply for industrial real estate, *Journal of the American Real Estate and Urban Economics Association*, 18(4), 530–47.

Wheaton, W. C., Torto, R. G. and Evans, P. (1997) The cyclic behaviour of the Greater London office market, *The Journal of Real Estate Economics and Finance*, 15(1), 77–92.

Worzala, E. (1994) Overseas property investments: how are they perceived by the institutional investor?, *Journal of Property Valuation and Investment*, 12(3), 31–47.

Worzala, E. (1995) Currency risk and international property investments, *Journal of Property Valuation and Investment*, 13(5), 23–38.

Worzala, E. M. and Bajtelsmit, V. L. (1997) Real estate asset allocation and the decision-making framework used by pension fund managers, *Journal of Real Estate Portfolio Management*, 3(1), 47–56.

Worzala, E. and Bernasek, A. (1996) European economic integration and commercial real estate markets: an analysis of trends in market determinants, *The Journal of Real Estate Research*, 11(2), 159–81.

Worzala, E. M., Johnson, R. D. and Lizieri, C. M. (1997) Currency swaps as a hedging technique for an international real estate investment, *Journal of Property Finance*, 8(2), 134–51.

Wyatt, P. (1996) The development of a property information system for valuation using a geographical information system (GIS), *Journal of Property Research*, 13(4), 317–36.

Young, M. S., Geltner, D. G., McIntosh, W. and Poutasses, D. M. (1995) Defining commercial property income and appreciation returns for comparability to stock market-based measures, *Real Estate Finance*, 12(2), 19–30.

Yusof, A. (1999) *Modelling the impact of depreciation: a hedonic analysis of offices in the City of Kuala Lumpur, Malaysia,* Unpublished PhD thesis, University of Aberdeen.

Ziobrowski, A. J. and Curcio, R. J. (1991) Diversification benefits of US real estate to foreign investors, *The Journal of Real Estate Research,* 6(2), 119–42.

Ziobrowski, A. J., McAlum, H. and Ziobrowski, B. J. (1996) Taxes and foreign real estate investment, *The Journal of Real Estate Research,* 11(2), 197–213.

Ziobrowski, A. J. and Ziobrowski, B. J. (1993) Hedging foreign investments in US real estate with currency options, *The Journal of Real Estate Research,* 8(1), 27–54.

Ziobrowski, A. J., Ziobrowski, B. J. and Rosenberg, S. (1997) Currency swaps and international real estate investment, *Real Estate Economics,* 25(2), 223–51.

Ziobrowski, B. J. and Ziobrowski, A. J. (1997) Higher real estate risk and mixed-asset portfolio performance, *Journal of Real Estate Portfolio Management,* 3(2), 107–15.

Index

9 780582 316126